Carolin Schurr
Performing Politics, Making Space

ERDKUNDLICHES WISSEN

Schriftenreihe für Forschung und Praxis

Begründet von Emil Meynen

Herausgegeben von Martin Coy, Anton Escher und Thomas Krings

Band 152

Carolin Schurr

Performing Politics, Making Space

A Visual Ethnography of Political Change in Ecuador

Franz Steiner Verlag

Interfeminas Förderbeitrag

Umschlagabbildung:
© Carolin Schurr: Wahlkampagne der Partei Pachakutik in der Stadt
Francisco de Orellana

Bibliografische Information der Deutschen Nationalbibliothek:
Die Deutsche Nationalbibliothek verzeichnet diese Publikation in der Deutschen
Nationalbibliografie; detaillierte bibliografische Daten sind im Internet über
<http://dnb.d-nb.de> abrufbar.

Dieses Werk einschließlich aller seiner Teile ist urheberrechtlich geschützt.
Jede Verwertung außerhalb der engen Grenzen des Urheberrechtsgesetzes
ist unzulässig und strafbar.
© Franz Steiner Verlag, Stuttgart 2013
Druck: Laupp & Göbel, Nehren
Gedruckt auf säurefreiem, alterungsbeständigem Papier.
Printed in Germany.
ISBN 978-3-515-10466-1

TABLE OF CONTENTS

List of Figures 9
Geographies of Thanks 11

DEPARTURES

Three Vignettes: Encountering Women in Electoral Politics 17

1. INTRODUCING POLITICAL CHANGE IN ECUADOR 19
Searching for a Feminist Electoral Geography 22
Puzzling over Political Transformations in Ecuador 27
Traveling through Ecuadorian Local Politics 35

2. RETHINKING ELECTORAL GEOGRAPHY 39
Theorising Spaces of Democracy 43
Understanding (Spaces of) Politics as Antagonistic 43
From the Performativity of Gender to Performative Spaces of Politics 49
The Emotional Spaces of Passionate Politics 53
(Spaces of) Politics at the 'Crossroad' of Gender, Ethnicity, Class and Locality 56
Rethinking Electoral Geography through Feminist Theories 61
Studying Political Practices in Local Spaces of Politics 63

PERFORMATIVITIES

3. THE PERFORMATIVITY OF POST-COLONIAL POLITICS 73
Taking Butler and Mouffe Elsewhere 73
Thinking Space through Performativity and Antagonism 74
Hegemony, Power and Spaces of Politics 77
Contesting (Post-)Colonial Political Hegemonies 82
Performing Antagonism in Chimborazo's Local Politics 84
Interculturalidad as Agonistic Politics 89
Towards a Political Geography of Change 93

4. A VISUAL ETHNOGRAPHY OF POLITICAL PERFORMANCES 95
Opening Credits 95
Genre: Visualizing the Performative Turn 96
Stage Direction: Struggling with the Camera in the Field 99
On the Set: Identity Performances on the Political Stage 100
Closing Credits 104

EMOTIONS

5. PERFORMATIVE EMOTIONS IN ELECTORAL CAMPAIGNS 109
Que viva Pachakutik: Emotions in Local Campaigns 109
The Performativity of Emotions in Political Speech 111
Researching Emotions 114
Emotional Geographies of Populism 116
Emotional Geographies of Local Campaigning 120
Towards an Emotional Electoral Geography 125

6. THE INTERSECTIONALITY OF EMOTIONS IN CAMPAIGNS 127
Bridging Emotional and Political Geographies 128
Researching Emotions from an Intersectional Perspective 129
Structural Intersectionalities as Cause and Target of Collective Feelings 130
Discourses around the Whiteness and Masculinity of Ecuador's Political Spaces 133
Emotional Performances as Part of Political Identity Constructions 134
Thinking Emotions from an Intersectional Perspective 136

INTERSECTIONALITIES

7. RETHINKING GENDER QUOTAS INTERSECTIONALLY 141
Accessing Ecuadorian Politics 141
Intersectionality in Electoral Politics 142
Researching Intersectionality in Electoral Geography 145
Gender Quotas as a Mechanism for Enriching Democratic Inclusion? 147
Intersectional Political Agendas? 151
A Geographical Perspective on Gender Quotas 164

8. TACKLING FEMINIST POSTCOLONIAL CRITIQUE THROUGH PARTICIPATORY AND INTERSECTIONAL APPROACHES 167
Facing Feminist Postcolonial Critique 167
Towards a Feminist Postcolonial Research Agenda 168
Participatory Approaches between Tyranny and Transformation 169
Taking into account Intersectionality 172
Developing Transnational Feminist Research Practices 176

CONCLUSIONS

9. TOWARDS FEMINIST ELECTORAL GEOGRAPHIES 181
(How) Do Women Do Politics? 181
Performing Politics 182
Making Space for Feminist Electoral Geographies 185

REFERENCES 189

LIST OF FIGURES

Figure 1: Diana Ataimant in the National Assembly 17
Figure 2: Leaflet of the National Women's Office 17
Figure 3: Matilde Hidalgo de Prócel 28
Figure 4: The Inti Raymi uprising 29
Figure 5: Campaign propaganda of Pachakutik 31
Figure 6: Leaflet of the National Council of Women 33
Figure 7: Women's share in political offices since 1979 34
Figure 8: Colonial painting in the municipality of Riobamba, Chimborazo 81
Figure 9: Cover of the political agenda of CONAIE 92
Figure 10: Indigenous leaders in the National Assembly 92
Figure 11: Electoral march 95
Figure 12: Emma waving 95
Figure 13: Constituents 95
Figure 14: Guadalupe Llori on the stage 101
Figure 15: Guadalupe Llori's campaign poster 102
Figure 16: Guadalupe Llori 103
Figure 17: Che Guevara 103
Figure 18: Fidel Castro 103
Figure 19: Guadalupe Llori's campaign speech 110
Figure 20: Crowd cheering 110
Figure 21: Bucaram 'bathing' in the crowd 119
Figure 22: Bucaram crucified 119
Figure 23: Bucaram crucified 119
Figure 24: Marco Santi 121
Figure 25: Crowd cheering 121
Figure 26: Marco Santi 123
Figure 27: Conceptual differences among approaches to the study of race, gender, class
and other categories of difference in political science 145
Figure 28: Word cloud 'Madre' 153
Figure 29: Mayor Anita Rivas in an event with the Association of Kichwa Women 154
Figure 30: Handicraft center at the wharf in Francisco de Orellana 154
Figure 31: Word cloud 'Maestra' 156
Figure 32: Prefect Lucia Sosa and her educational project 157

Figure 33: Word cloud 'Feminista' 159
Figure 34: Juridical women's office in Riobamba 161
Figure 35: Indigenous women demanding 'we want a life without violence' 162
Figure 36: Do women do politics? 181

LIST OF MAPS

Map 1: Participation of women in electoral office in the three case study areas 65
Map 2: Women in executive and legislative positions in Ecuadorian local politics 149

LIST OF TABLES

Table 1: Comparing the research agenda of conventional and new electoral geographies 43
Table 2: Approaching feminist electoral geographies through antagonism 48
Table 3: Approaching feminist electoral geographies through performativity 52
Table 4: Approaching feminist electoral geographies through emotions 56
Table 5: Approaching feminist electoral geographies through intersectionality 60
Table 6: Information on case study localities and interviews conducted 66
Table 7: Suggested methodological research agenda for a feminist electoral geography 68
Table 8: Women in executive and legislative positions in Ecuadorian local politics 148
Table 9: Ethnic and gender composition of the Juntas Parroquiales Rurales 150
Table 10: Proposed approach for a feminist electoral geography 187

GEOGRAPHIES OF THANKS

> 'The academy is not paradise. But learning is a place where paradise can be created' (hooks 1994: 207).

Feminist scholars have long insisted that research is always a collaborative endeavor. It is my heartfelt wish to acknowledge the support, help, inspiration, and collaboration of those people who have significantly contributed to making this research project happen.

My interest in Latin American politics began at the start of the new millennium in London, where a Brazilian friend convinced me to backpack with her through Latin America. Travelling up the Brazilian coast, we discussed with the *Movimento dos Trabalhadores Rurais Sem Terra* questions of land rights; on board of a ship along the Amazon, we had passionate conversations about the rights of indigenous people; in Bolivia, we became excited about the emerging power of the indigenous movement; in Argentina, we were confronted with the current economic crisis, when we ended up in the middle of riots in front of a bank. This long journey through a number of Latin American countries piqued my interest to learn more about this colorful continent full of contrasts. *Obrigada a* Lou for introducing me to this wonderful part of the world.

Hoping to better understand this 'New World', I started studying Geography and Latin American Studies in Eichstätt. A student exchange program of the DAAD offered me the possibility to study at the Pontificia Universidad Católica del Ecuador in Quito. At the same time, I worked as an intern at the German Development Cooperation (GIZ). While I learned much about Luhmann's System Theory, Beck's Risikogesellschaft, and Marx's Kapital in my classes, it was through the internship and my manifold travels to the province of Cotopaxi that I came to better understand the power relations that saturate Ecuadorian politics. I was lucky to have a very critical group of colleagues at the GIZ that not only introduced me to critical development studies, post-development, and postcolonial theory but also to the particularities of Andean life through sharing their everyday lives with me. *Miles de gracias a* Oscar Forrero for questioning my understanding of development, to Edison Mafla, María Belén Molina, Marjorie Reinoso, Sofía Starnfeld for the *fiestas* and *cervezas*, and especially to Lorena Cantillo for her love, her hospitality, and the many '*escapadas*' that distracted me in wonderful ways from time to time from my 'real' work in Ecuador.

Back in Germany, it was Prof. Hans Hopfinger and Marc Boeckler who nourished my passion for academic research in the first place and who encouraged me to delve into Latin American politics in my Master thesis. Hence, *ein großes Dankeschön* for introducing me to the academic world.

Returning to Ecuador to conduct my PhD research in 2008, I further received valuable support from the Facultad Latinoamericana de Ciencias Sociales (FLACSO) in Quito. So, *gracias* a Alba Di Filippo, Gioconda Herrera, Violeta

Mosquera, and Carlos de la Torre for offering me an inspiring academic environment far from my home university. The biggest *gracias* goes to the women politicians in Esmeraldas, Orellana and Chimborazo who were so kind to open the doors of the provincial councils, municipalities, and local governments for me and gave me an insight in the everyday routines of local politics in Ecuador. Further, *gracias* to the Arias Cardenas 'clan' for all their love and support, especially for offering me to stay in their house in Esmeraldas while carrying out research in this province.

A number of friends and colleagues read chapters of this book at various stages of draft form. These include my supervisor Lynn Staeheli, Katharina Abdo, Jonathan Everts, Bettina Fredrich, Benedikt Korf, Juliet Fall as well as a number of anonymous reviewers. An international *thanks* for their time, encouragement, and thoughtful critique. Another *Dankeschön* goes to Dörte Segebart who co-authored one of the chapters of this book.

In Bern, where I was based during the years of my PhD, a bunch of colleagues, both at the Department of Geography and the Interdisciplinary Center for Gender Studies (IZFG), have substantially inspired this research. *Merci* to Bettina Fredrich, Yvonne Riaño, Patricia Felber, Renate Ruhne, Jenny de la Torre, Brigitte Schnegg, Andrea Kofler, Fabienne Amlinger, and all the other 'Gradis'. Another *merci* goes to Doris Wastl-Walter who has supervised this work. I have been helped by excellent research assistants. I thank Urezza Caviezel, Dina Spörri, Lukas Erhard, and Ophélie Ivombo. Susanne Henkel and Sarah Schäfer from the Franz Steiner Verlag have guided me in a very supportive way through the publication process.

I would like to acknowledge the financial support of the Swiss National Science Foundation and the Department of Geography at the University of Bern. I am especially thankful for the support received in the Graduate School 'Gender: Scripts and Transcripts' also financed by the Swiss National Science Foundation. With a generous financial contribution, the foundation Interfeminas made the publication of my research possible.

Some of the material of this book has been published elsewhere. I thank *Geographica Helvetica* for permission to reprint the following two papers in slightly revised form in chapter 4 and 8: Schurr, Carolin, 2012. Visual ethnography for performative geographies: How women politicians perform identities on Ecuadorian political stages. *Geographica Helvetica* 67 (4), 195–202 and Schurr, Carolin, Segebart, Dörte, 2012. Tackling feminist postcolonial critique through participatory and intersectional approaches. *Geographica Helvetica* 67 (3), 147–154. An earlier draft of chapter 3 has been published as Schurr, Carolin, 2013. Performativity and Antagonism as Keystones for a Political Geography of Change, in: Glass, Michael, Rose-Redwood, Reuben (Eds.), *Performativity, Politics and Social Space*. Routledge, New York. An earlier version of chapter 5 has appeared as Schurr, Carolin, 2013. Towards an emotional electoral geography: The performativity of emotions in electoral campaigning in Ecuador. Geoforum 49, 114-126. Chapter 6 has been published in a Portuguese translation as Schurr, Carolin, 2012. Pensando emoções a partir de uma perspectiva interseccional: as geografias emocionais das campanhas eleitorais equatorianas. Revista Latino-Americana de Geografia e Gênero. Revista Latino-Americana de Geografia e Gênero 3 (2), 3–15. Unless stated otherwise, the copyright of the images, tables, maps and graphs lie with the author. I have made every attempt to contact the institutions and people who hold the copyrights of the images and graphs I have used. The institutional landscape in Ecuador is characterized by fast changes that make it difficult to identify copyright holders. The fact that I finish this book some thousands miles away from Ecuador makes this endeavor impossible. I hope the copyright holders agree with the way I have used their images.

Finally, *ein herzliches Dankeschön* to Antoinette who taught me the importance of finding my own rhythm, to my dearest friends for showing me that there is also a life outside of academia, to my family for waiting patiently for me to finish this work even though it must have seemed strange to them that my little niece learned to walk, talk, and sing, while I wrote 'just' 200 pages. I wish I had finished this book while letters and words still made sense to my father. I am very fortunate that my mother with her endless interest for the world has always made big efforts to read, understand, and discuss my work. The last and *herzlichste Dankeschön* goes to Martin for being in my life, engaging with my thoughts, and turning every day into a sunny day.

DEPARTURES

THREE VIGNETTES:
ENCOUNTERING WOMEN IN ELECTORAL POLITICS

September 18th, 2005,
car ride on a stony road to Corazon Grande, Ecuador

Radio: In Germany, Angela Merkel has just been elected the first woman Chancellor.

Lorena: Wow, you gonna be governed by a woman! What is this going to be like?

Me: I don't know, I can't imagine it. I have been governed nearly all my life by this big man, Helmut Kohl.

March 3rd, 2008, national congress Ecuador
Interview National Asambleista Diana Ataimant:

"This political space I am occupying is a space from which women and especially Shuar women have always been excluded."

Figure 1: Diana Ataimant in the National Assembly

April 14th, 2009, Provincial Council of Orellana
Interview Prefect Guadalupe Llori:

"We didn't have a Comisaría de la Mujer [legal office where women can report sexual violence], so we created the Comisaría de la Mujer. Back then I was the Mayor and we build this shelter, because this was a pueblo machista who violated the women."

Figure 2: Leaflet of the National Women's Office

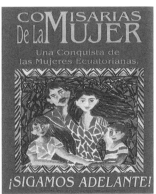

1.
INTRODUCING POLITICAL CHANGE IN ECUADOR

The huge white body of Helmut Kohl epitomized and embodied politics throughout my childhood and adolescence. As the vignettes show, it was about the time when Angela Merkel became the first female Chancellor of Germany, when my interest in the co-constructive processes of gendered and ethnic identities and political spaces in the Ecuadorian context emerged. Working as a graduate student with local politicians in the highland province of Cotopaxi, I became fascinated with the transformations occurring at fast pace in Ecuadorian politics (Schurr 2009a). In 2004, when I first started to conduct research in Ecuador, the success of social movement struggles became more and more visible: Indigenous movements had successfully established indigenous people[1] as political subjects as a consequence of the overthrow of President Jamil Mahuad by indigenous mass protest (O'Connor 2003, Selverston-Scher 2001, Van Cott 2008, Yashar 2006b, Zamosc 2004). Within the indigenous movement, indigenous women like Diana Ataimant, cited above, increasingly gained space in electoral politics with Nina Pacari appointed to the post of foreign minister as the first indigenous woman in 2003 (Andolina, Laurie, and Radcliffe 2009, Pacari 2005, Prieto et al. 2006, Radcliffe 2002). In the elections of 2004, the women's movement managed to legally penalize all political parties whose electoral lists did not conform to conditions of the gender quota law, which required the parties to alternate men and women in equal numbers in the electoral lists (Quezada 2009, Vega Ugalde 2005). As a result, women presented over 40 percent of the candidates of the 2004 elections. At the beginning of the new millennium, women, indigenous and Afro-Ecuadorian people became elected as mayors and prefects for the first time as a result of successful struggles of both ethnic and women's movements (Arboleda 1993, Lalander 2010, Radcliffe, Laurie, and Andolina 2002, Van Cott 2008). Taking into account that most female, indigenous, Afro-Ecuadorian peasants and workers were denied their political citizenship rights until 1979, when literacy requirements for suffrage were eliminated, these are stunning developments. As

---

1 The term '*indigena*', introduced by the Spanish colonizers, reflects the power those have who name and define other people. Further, the term homogenizes the numerous indigenous ethnicities; alone in Ecuador, indigenous people identify with fourteen different *nacionalidades* y sixteen pueblos. 'Indigenousness', however, has changed over time and has been reconnoted positively by the indigenous movement (Radcliffe 1997).

women, indigenous and Afro-Ecuadorian people first gained ground in the spaces of local politics such as rural parishes, municipalities, and provincial councils, I was eager to learn more about the way ethnic, gender and class differences were negotiated between traditional and new political subjects in local politics. The term 'new political subjects' is used here to refer exactly to these emerging political subjects – mainly women, indigenous and Afro-Ecuadorian people – who have been excluded from formal citizenship since colonial times. While these social groups have a long history of political struggle in Ecuador (Prieto 2004, Prieto and Goetschel 2008), they have appeared in institutionalized politics only recently in more significant numbers. The term new political subjects results from their participation in 'new social movements' through which they have fought for their political rights (Alvarez, Dagnino, and Escobar 1998, Escobar and Alvarez 1992).

Processes of political transformation have been at the center of my personal and academic concern for a while, as the vignettes above show. What has really intrigued me, in the change of power to Angela Merkel, Diana Atainmant or Guadalupe Llori and all the other women I have encountered in my research, is the role identity and difference play in the construction of political spaces in postcolonial contexts. While I have still just a very partial answer to the question whether women make a difference in electoral politics (and vignette number 3 certainly shows that they do), I have been surprised by the diverse motivations, struggles, competencies, knowledges, thoughts, convictions and emotions of the women politicians I encountered during my research. I was struck by the differences I found among women politicians, but also the similarities of their daily struggles they face beside their diverse (political) biographies. These constant tensions between differences and commonalities that characterize women's experience in electoral politics, however, are barely acknowledged in the popular and academic writing about women in politics despite a fast growing body of literature dealing with women's participation in electoral politics. The media is more interested whether women are actually the better (read: less corrupt and more beautiful) politicians (Lüneborg 2009), international organizations are more concerned about the rise of women into electoral office to make progress towards the third Millennium Development Goal (BMZ 2007, Byanyima 2007), and academic literature still focuses primarily on gender as a category of difference that structures the spaces of politics, by and large not taking into account the intersectionality of gender, ethnicity, class, sexuality, and locality (Craske 2003, Lovenduski 2005, 2010, Sauer 2008). Being sensitive to the intersectionality of social structures and identities that shape spaces of politics and having in mind their time-spatial situatedness, the aim of this study is to fill the blank spots at the intersections of gender, ethnicity, class and locality in the way participation in electoral politics is performed, experienced, felt, thought, depicted, and tackled.

The reasons that have driven the making of this research are essentially three: First, women have been increasingly successful in winning elections on all political levels in Latin America. Of the 33 countries in the Latin American and Caribbean region, nine have elected female presidents or prime ministers, an achievement unparalleled elsewhere in the developing world. Michele Bachelete (Chile

2006–2010), Cristina Fernández de Kirchner (Argentina 2007–2011), and Dilma Rousseff (Brazil 2011–2015) are just the most prominent examples. While much has been written about women in national politics (e.g. Bush 2011, Escobar-Lemmon and Taylor-Robinson 2005, James 1997), little is known about the women who have entered local politics as a result of gender quota laws implemented throughout the region between 1997–2002 (Peschard 2003). At the same time, indigenous movement struggles have resulted in the institutionalization of indigenous politics in form of ethnic political movements and the election of indigenous politicians, most famously Evo Morales in Bolivia (Andolina 1999, Becker 2008b, Lucero 2008, Rice 2011, Rice and Van Cott 2006, Selverston-Scher 2001, Van Cott 2000, 2006, 2008, Yashar 1999, 2006a, 2006b). While the indigenous and women's movement have been investigated mostly on separate terms or by focusing explicitly on the political agency of indigenous women, I am interested in the intersections between the two movements. Hence, this research focuses on the interplay between gender and ethnicity in transformation processes in electoral local politics.

Second, to foster social change and enrich democracy, we need to know more about the way new political subjects like women, indigenous and Afro-Ecuadorian people shape the constitution of both the discursive and material spaces of politics in post-colonial contexts[2]. Focusing on the everyday practices of these so called new political subjects, I aim to question how the presence of new political subjects actually transforms political agendas, alters the constitution of political spatialities and renegotiates access to the spaces of politics. In so doing, I bring to the center of attention democratic processes in post-colonial societies that have been neglected by an electoral geography that has mostly focused on the core countries in the Western hemisphere (Flint and Taylor 2007: 195).

Third, by focusing on the way new political subjects constitute and transform the very spatialities of politics, I would like to go beyond existing research in electoral geography and political science about women's political representation that is satisfied with presenting the number of women in electoral politics and discussing the gendered effect of policies launched by women. As this work builds upon these studies, it is far from arguing that this knowledge is not useful, but rather wants to suggest that there is the necessity to move further, to delve into the messiness, contradictions and affective dimensions of local politicians' everyday lives. While much has been written on the structural, institutional and cultural bar-

---

2   Throughout my work, I differentiate between post-colonial and postcolonial. Sharp (2009a: 3–5) has pointed out the importance of the hyphen in differentiating between the post-colonial as the period following independence from colonial powers and postcolonial as a critical approach that challenges colonialism and the values and meanings it depended upon (for a problematisation of the term see further Appiah 1991, Hall 1996a, McClintock 1995).

riers that women face on their way to electoral offices (for an overview see Norris and Inglehart 2001), too little has been said about the way political subjectivities and spatialities are made through the everyday practices and performances on the local political stages like the plaza where a campaign event takes place, the saloon of the municipality where the town council meets or a school which is inaugurated by the mayor. This book seeks to shed light on the way political subjectivities are produced and reproduced in the daily encounters on these diverse political 'stages' by asking: How are the subjectivities of politicians and the spatialities they bring into being gendered, racialized, ethnicized and classed through the practices, performances and interactions between the politicians, their audience and the time-spatial context they are embedded in? And, how do these practices, performances and interactions contribute to social change and processes of decolonization?

SEARCHING FOR A FEMINIST ELECTORAL GEOGRAPHY

Being immersed in the messy everyday business of local politics in Ecuador, I came to understand that electoral geographies are far more than electoral outcomes, which political and electoral scientists – including electoral geographers – preferably deal with when engaging with issues about the gendering or ethnicization of politics. To understand the processes of political change in Ecuador, for example, it was crucial to look at the relation between electoral and social movement politics rather than to stop investigating at the doorsteps of political institutions. To grasp the notion and meaning of gendered or ethnic identity performances in electoral campaigns, it was necessary to look back at the (post-)colonial history of political citizenship in Ecuador. To sense why citizens were so excited about the new president Rafael Correa, one had to be immersed in the crowds cheering at him. These three little empirical examples identify some of the key issues of an electoral geography that goes beyond electoral processes and results. As this book will show, antagonistic relations between different political communities, everyday performances of intersecting identities, and emotions[3] all play a crucial role in the construction of spaces of institutionalized politics.

---

[3] Recent discussions in human geography differentiate between emotions as social constructions and affects as direct bodily, pre-cognitive, biological forces (see Pile 2010, Thien 2005, Thrift 2009). I use these two terms synonymously arguing that the differentiation (and dichotomization) between emotions and affects does not recognize that 'even seeming direct responses actually evoke past histories, and that this process bypasses consciousness, through bodily memories. So sensations may not be about conscious recognition, but this does not mean they are "direct" in the sense of immediate.' (Ahmed 2004a: 39).

Despite the fact that the analysis in this book goes far beyond the study of electoral results, I still situate the book in the subfield of electoral geography. The reasons for this are twofold. First, the book focuses on the gendering of institutionalized politics and elections. While elections and national politics have for a long time been the 'heart' or bread-and-butter business of political geography, feminist political geographers have called (successfully) for the need to open up the narrow, masculinist, and state-centered perspective of mainstream political geography (see e.g. Dowler and Sharp 2001, Hyndman 2004, Kofman 2008, Kofman and Peake 1990, Schurr and Fredrich 2011, Staeheli, Kofman, and Peake 2004). Challenging the gendered binaries about the key analytical categories of political geography such as private/public spaces, formal/informal politics, reason/emotion (Brown and Staeheli 2003, Brownill and Halford 1990, England 2003, Fincher 2004a, Sharp 2003, Staeheli 1996, Staeheli and Mitchell 2004), feminist political geographers have argued that

> 'the political is not just relevant to elections, the state and the international conflict, [but] it is seen in the ways in which women mobilize at the grass roots, in the ways an ethic of care is brought into political discourse, in the ways masculinity and femininity are invoked in ideas of nation and international conflicts' (Staeheli and Kofman 2004b: 6).

In an attempt to acknowledge the call of feminist political geographers to expand the boundaries of political geography beyond issues of elections and state politics, it would be counter-productive from a feminist perspective to reduce the wide field of political geography again to elections and institutionalized politics (even if understood in a broad sense). Hence, rather than situating my study in political geography, the book develops what I call a feminist electoral geography in an effort to recognize the importance of not conflating political geography with (and constraining it to) electoral geography. It rethinks electoral geographies by sketching new ways to approach electoral geographies theoretically through theories of antagonism, performativity, and intersectionality, empirically through focusing on the local, the body and emotions, as well as methodologically through (visual) ethnographies and feminist postcolonial approaches.

Second, developing explicitly a feminist *electoral* geography, I argue that feminist political geographies have neglected institutionalized and electoral politics in their attempt to refocus attention on diverse political settings beyond the state, such as social movement politics (Conway 2008, Gruszczynska 2009) or politics of care (England 2003, Pratt 2004). While feminist political geographies' broadening of what gets counted as political subject matter is a positive move, it has unfortunately, according to Barnett and Low (2004), led to a problematic rejection of what is seen as 'ordinary' political subject matter, such as elections and political parties. This rejection not only 'runs the risk of jettisoning any concerns for the realms in which politics most obviously still goes on' (Barnett and Low 2004: 6), but also misses the fact that institutionalized politics are one of many sites where decisions are taken that shape women and men's lives in different ways. Hence, we need to recognize that feminist political geographies must refocus attention on the gendered dimension of institutionalized politics and elections.

In my attempt to develop a feminist electoral geography that deals with the messiness of everyday political life in post-colonial Ecuador, I have been inspired by different theoretical strands that electoral geographers have not yet engaged with and which I would like to introduce to electoral geography and electoral studies in general. In so doing, the book aims to contribute to broader debates in political geography by developing a (feminist) electoral geography that is inspired and builds on geography's recent turn(s)[4] towards practices and performativity, the body and embodiment, and affect and emotion. Political geographies have engaged to different extents with these new turns and I therefore argue that political and electoral geographies can equally benefit from a fuller engagement with the conceptual implications of these turns.

The performative turn can be understood as direct response to the focus on texts and representations that has dominated new cultural geographies by shifting attention to the performances and practices of everyday live (Boeckler and Strüver 2011, Dirksmeier and Helbrecht 2008, Gregson and Rose 2000, McCormack 2009, Nash 2000, Pratt 2004, Strüver and Wucherpfennig 2009, Thrift 1997, Thrift 2003). Scholars in political geography and critical geopolitics have for quite a while now criticized their disciplines for their 'mesmerized attention to texts and images' (Thrift 2000: 381), advocating a 'critical geopolitics that is more attuned to everyday practices' (Müller 2008: 329) and 'everyday-life geopolitics' (Paasi 2006: 217). Nick Megoran's (2005, 2006) ethnographic work on the impact of the partial closure in 1999–2000 of the Uzbekistan-Kyrgyzstan Ferghana Valley boundary or Sara Koopman's (2011) long-time ethnography of international peace accompaniers in Colombia are excellent examples of how the practice turn has been integrated in political geography and critical geopolitics. In electoral geography, however, such an engagement with the everyday practices of campaigning, canvassing, voting, organizing and governing has scarcely taken place.

The focus on practices has implied an increasing interest in processes of embodiment and the body itself (Colls 2007, 2012, Harrison 2000, Longhurst 1997, 2001, Mahtani 2002, McDowell 2009, Nelson 1999b, Simonsen 2013, Slocum 2008, Strüver 2005a). Feminist geopolitics and feminist political geography have been at the forefront of thinking about the embodiment of political actions and the role of bodies in the construction of (geo-)political spaces such as the nation (Faria 2013, Marston 1990, Mayer 2004, Radcliffe 1996, 2000, Radcliffe and Westwood 1996). In their seminal paper 'A feminist geopolitics', Dowler and Sharp (2001: 169) advocate

---

4   While the performative turn, the practice turn, the affective turn are often referred to as different turns, they can all be considered as reactions and responses to new cultural geographies' focus on texts and representations (for an overview over the critique see Thrift and Dewsbury 2000).

'recognising the inherent and unavoidable embodiment of geographical processes and geopolitical relationships at different scales. In order to rewrite the everyday experiences of individuals back into geopolitical events, academics are relating the scale of their investigations from the global and national to that of the community, home and body'.

Jennifer Fluri's (2011a, 2011b) work on the embodied geopolitics of the recent political conflict in Afghanistan is exemplary for a feminist geopolitics that redeems Dowler and Sharp's call. Fluri highlights the corporeal as a key site of analysis for the everyday and seemingly apolitical spaces occupied by civilians living amidst political conflict. The work of Sara Smith (2011, 2012) and Banu Gökarıksel (2009, 2011) show how women's bodies are turned into sites of political struggle. Smith's 'intimate geopolitics' discusses how in the Leh District of India's Jammu and Kashmir State, political conflict between Buddhists and Muslims has been articulated in part through women's bodies by preventing inter-religious marriages. In a similar vein, Gökarıksel considers women's headscarves in Turkey as objects of political struggles and shifting embodied expressions of political ideologies. The feminist electoral geography I seek to develop builds on this body of literature in feminist political geography and feminist geopolitics, asking how bodies matter in the construction of electoral spaces and how differently gendered, racialized, ethnicized and classed bodies matter in different ways in particular political sites.

Despite the recent boom in geographies of emotion and affect (Anderson and Smith 2001, Bondi, Davidson, and Smith 2005, Davidson, Bondi, and Smith 2005, McCormack 2006, Pile 2010, Sharp 2009b, Smith et al. 2009b, Thien 2005, 2011, Tolia-Kelly 2006, Woodward 2011), Pain et al. (2010: 973) argue that emotional geographies have been oriented 'more towards social, cultural and environmental dimensions, than mapping out the political geographies of emotion'. Still, this book has benefited immensely from insights of emerging work under the label of 'emotional geopolitics' (Dodds and Kirby 2012, Katz 2007, Pain 2009, 2010, Pain et al. 2010, Pain and Smith 2008, Wright 2008) and 'politics of affect' (Barnett 2008, McCormack 2006, Thrift 2004, 2009). On the one side, 'emotional geopolitics' have focused my attention to the way 'emotions [are] experienced as simultaneously *both* local *and* global' (Pain 2009: 476) and the need to 'incorporate emotions in nuanced and grounded ways' (Pain 2009: 474) in political analysis. Translating these claims into electoral geography has meant for me to question how feelings of citizens in local events are connected and related to national and international current and historical events, analyzing the encountered emotions on different scales. The second strand that has informed my thinking consists of discussions taking place in non-representational geographies about the politics of affect. Non-representational approaches taught me to pay more attention to the 'systematic engineering of affect' (Thrift 2004) in electoral politics. While I critically question many of the assumptions of this body of work, such as its lacking attention to social and bodily differences (Colls 2012, Thien 2005, Tolia-Kelly 2006) or the primacy given to the unconscious (Bondi 2005a, Korf forthcoming, Schurr forthcoming), I found it rewarding to sense with my own body the affective capacities in the spaces between bodies in political collectives beyond linguistic articulations (McCormack 2003, Roelvink 2010).

Inspired by these three turns, I sketch in this book a feminist electoral geography that (1) understands the production of political spaces as the outcome of antagonistic practices and embodied performances, (2) questions the role emotions play in the constitution of political spaces, and (3) focuses on the way markers of difference such as gender, race, ethnicity, class and so on intersect in political spaces and are constantly (re-)produced, stabilized, challenged and transformed through political practices. This conceptual framework has emerged from and at the same time frames my empirical engagement with local spaces of electoral politics in Ecuador. Studying processes of political change in Ecuador from a feminist perspective, the book aims to challenge the positivist, Eurocentric and masculinist framework of electoral geography (for a critique see Agnew 1990, 1996, Cupples 2009, Johnston and Pattie 2004, Secor 2004).

John Agnew's (1996) claim that 'context counts in electoral geography' guides this book in two ways. On the one hand, my conceptual thinking emerges from and is closely linked to a certain context: Ecuadorian politics. The context I studied, however, was not limited to the national borders of a physical territory. Rather, I follow Agnew (1996: 130) in using the term 'context' as a 'possibility of avoiding a specific scale commitment', understanding political actions as the outcome of influences emanating from a range of interrelated geographical scales. My empirical analysis engages with and interrelates scales from the body to the global, zooming in and out from a bodily performance to the global history of colonialism and back in again to the feelings of a local community. While my empirical fieldwork has focused on the context of Ecuadorian *local* politics, many developments in the municipalities can not be understood without 'zooming out' to *national* politics or the action of international political or development organizations or 'zooming in' to intimate relationships, emotional encounters or everyday place-specific routines. Different scales of zooming have demanded different forms of engagement with literature, concepts and theories. The process of knowledge production of this book was a constant back and forth from the 'field' to my desk in Ecuadorian offices and libraries, to the field, to the desk back home and so forth. In this vein, empirical data and conceptual debates have directly engaged with and shaped each other. The structure of the book represents this co-constructive relationship between empirical claims and conceptual thinking in so far as each chapter both makes conceptual and empirical claims, discussing theoretical aspects as well as presenting empirical data. On the other hand, I argue that not only the context in which empirical data is collected counts, but also the researcher's biographical context and his/her positionality. In so doing, I advocate a feminist standpoint theory that recognizes the partiality and situatedness of knowledge production – including the development of theoretical concepts (Harding 1991, Hartsock 1983). My positionality – both in the sense of my national, racial, gender, sexual, class position (Kobayashi 1994, Rose 1997b) and in the sense of my own (academic) biography regarding theoretical debates I have been engaging with and the national contexts I have been studying in – shapes the way I engage with (Ecuadorian) electoral geographies.

Following the claim that 'context counts', the introduction into the context of Ecuadorian politics (Puzzling over Political Transformations in Ecuador) precedes the conceptual introduction into (feminist) electoral geographies (Rethinking Electoral Geography). It is a very partial and situated account, emphasizing those political events that from my point of view (and standpoint) are crucial to understanding the recent political changes in Ecuador.

PUZZLING OVER POLITICAL TRANSFORMATIONS IN ECUADOR

'Latin American states are structured around hierarchical intersecting relations of class, race and gender with "first class" citizenship reserved for white, male, urban and wealthier individuals and "second class citizenship" for indigenous peoples, individuals of Afro-descent, women, and rural and poorer groups' (Radcliffe and Pequeño 2010: 984).

'Participation shall be governed by the principles of equality, autonomy, public deliberation, respect for differences, monitoring by the public, solidarity and interculturalism. [...] For multi-person elections, the law shall establish an electoral system in line with the principles of proportionality, equality of vote, equity, parity and rotation of power between women and men' (Constitución de la República del Ecuador 2008: Art. 95, Art, 116).

Reading the first quote from Sarah Radcliffe and Andrea Pequeño (2010) against the two articles of Ecuador's Constitution of 2008, one is puzzled by the discrepancy between the hierarchies observed with regard to political citizenship in Ecuador and the constitutional attempts to overcome these (post-)colonial gendered, racialized and classed hierarchies. It was this discrepancy between political reality and a progressive inclusive legislation that attracted me to engage with the recent processes of political transformation in Ecuador. I first came across this discrepancy when working with rural local governments in the Province of Cotopaxi (Schurr 2009a). Local women politicians told me in the same breath about the gender quota law and the discrimination they face nevertheless in their political everyday lives.

This discrepancy between a persistent exclusionary political culture and a progressive inclusionary legislation can only be understood against the backdrop of Ecuador's (post-)colonial history and recent political transformations resulting from social movement struggles. To grasp the messy relationship between (post-)colonial political structures and the dynamics of change that shape contemporary politics in Ecuador, one needs to engage with the 'entangled histories' (Randeria 2002) of Ecuador's political spaces. Radcliffe and Westwood's (1996) extensive reflection offers interesting insights into how the Ecuadorian (post-)colonial history and the nation that emerged out of this history are gendered, sexualized, racialized and classed in multiple ways:

'Ecuador embarked on a self-conscious nation-building process following independence from Gran Colombia in 1830. Although no longer under colonial rule the administrative separation of "Indians" and "Spaniards" continued into the republic, the imagined superiority of the cultured *raza* or race of Spaniards and their descendants rested upon the pre-eminence of the Spanish language and religion, and a shared sense of commonality with the other elite creole in neighboring countries. Populations racialized through the use of terms such as Indians, black, cholos (urbanized indigenous groups), *montuvios* (people of mixed heritage) and *mestizos* were excluded from national life during the early republican period. Indeed, the elite's dominance depended in part upon the reproduction of their difference from indigenous and mixed groups' (Radcliffe and Westwood 1996: 5).

While referring the reader to the extensive literature on issues of nation building in Ecuador (Adelman 2006, Ayala Mora 2008, Becker 2008a, Benavides 2004, Canessa 2005, O'Connor 2007, Prieto and Herrera 2007, Radcliffe 1996), here, I would like to sketch only a few selective moments and events of Ecuador's political history that are crucial to understanding how these gendered, racialized, ethnicized and classed order has been challenged over the past decades. I do so along four images that characterize each of these moments.

Figure 3: Matilde Hidalgo de Prócel (Source: Goetschel et al. 2007: 16)

1924: Women's suffrage

The story to be told in my work starts on the 10th of May in 1924 in Machala, in the South of Ecuador. On the grounds that the Constitution did not explicitly exclude women from the electorate, Matilde Hidalgo de Prócel was the first woman in Latin America to register to vote and later on in 1941 to be elected for a political office (Prieto and Goetschel 2008: 300). This event is important for the development of my story for two reasons: First, Matilde Hidalgo challenged the gendering of electoral politics as an exclusive masculine domain through her heroic act. In order to claim her political citizenship rights she had successfully fought for, Matilde Hidalgo crossed the gendered boundaries that formerly assigned the spaces of electoral politics exclusively to men and positioned women outside of these spaces like the polling station, the parliament or the municipality.

Second, this little episode shows through the body and biography of Matilde Hidalgo that gender is not the only category that structures the spaces of electoral politics. Matilde Hidalgo was only able to fight for her citizenship rights because of her racial, class, and local positioning.

While coming from a modest family due to her father's early death, her whitened *mestizo*[5] ethnicity made her eligible for obtaining education in a Catholic monastery that facilitated her access to university later on. Taking into account that literacy requirements for suffrage were only eliminated in the Constitution of 1979, her educational background was crucial for obtaining political citizenship rights. Hence, Matilde Hidalgo's story is exemplary of the way gender, class, and ethnicity intersect within the spaces of politics and how these spaces are constituted on the basis of these intersectional mechanisms of exclusion at the same time. Although thanks to Matilde Hidalgo Ecuador became the first country in Lain America to give women the vote, women's suffrage did not become mandatory until 1967 and especially indigenous, Afro-Ecuadorian and peasant women did not gain the right to vote until 1979 due to their lacking literacy (Quezada 2009, Radcliffe 2008b). In sum, it took women a long time to challenge the masculine hegemony characterizing the post-colonial spaces of electoral politics.

1990: The *Inti Raymi* uprising

Figure 4: The Inti Raymi uprising (Source: http://pelusaradical.blogspot.com/2010/06/)

On June 4, 1990, the National Indigenous Confederation, CONAIE, launched a nationwide *levantamiento* (uprising) – often called the '*levantamiento indígena de Inti Raymi*' because the uprising took place just before the traditional June solstice 'Sun Festival' (Inti Raymi) celebrations – that shut down the country for a week by blocking roads (for detailed information about the uprising see Lucas 2000, Sawyer 1997, Whitten, Scott Whitten, and Chango 1997). While indigenous people have a long history of resistance and protest against colonial and post-colonial exploitation, discrimination and injustices

---

5 The term *mestizo* is an 'ethnic' rather than a 'racial' category, marked by language, dress, and class status. The national project of mestizaje celebrated racial and cultural mixtures as a way of forging a unified and homogeneous nation (Safa 2005). The explicit hierarchization of identities is what makes it problematic.

(Lyons 2002, Pallares 2002, Prieto 2004, Radcliffe 2000), it was this uprising, that 'thrust Indigenous activists as political actors onto the national stage' (Becker 2008a: 175). Linking the uprising to the 500 years of resistance (making reference to the quincentennial of Columbus's voyage to the Americas taking place in 1992), the CONAIE-led indigenous movement presented the government with a list of sixteen points that revolved around cultural issues such as bilingual education, economic concerns including access to credit and budget for development programs in indigenous communities, as well as political demands. The central political claims turned around the ending of central political control over local communities and amending the first article of the constitution to declare Ecuador to be a pluri-national and multicultural state (Pallares 2002: 228). For the indigenous movement, the 'pluri-national' signifies 'the recognition of a multicultural society in the indissoluble unit of a state that recognizes, respects and promotes unity, equality and solidarity among all peoples and nationalities, regardless of their historical, political and cultural differences' (Walsh 2009: 176). Identifying themselves as distinct 'nationalities' rather than an ethnic group or ethnic minority and calling Ecuador a 'pluri-national' country were strategic steps for the indigenous movements. In so doing, they not only aimed to achieve public recognition of specific indigenous rights as collective rights as ancestral groups who had first occupied and governed the land. As Pallares (2007: 149) shows,

> 'giving them status as nationalities, they hoped, would not only differentiate them from other socially subordinate groups with claims, including blacks and most coastal peasants, but would assign them a special place at the negotiating table with state officials and non-indigenous political actors'.

What is crucial about the uprising and the related claims for a pluri-national country is Becker's observation that indigenous people and the movement itself turned into a political actor through this massive uprising. Certainly, it is rather a whole series of protests, marches and events taking place during the 1990s through which indigenous subjects gained political power (Andolina 1999, Selverston-Scher 2001, Van Cott 2006, Zamosc 2004). The *Inti Raymi* uprising, however, can nevertheless be read as a first step of the indigenous movement to blur the boundaries between non-institutionalized and institutionalized politics. Through the direct negotiation of their demands with the government, the movement positioned itself as a political actor on equal footing with the government. While the protest took place mainly in public spaces like streets and plazas, to present their demands to the government of Rodrigo Borja a group of indigenous representatives entered the national congress. The march literally forced the predominantly white and *mestizo* government to open the doors of the national congress to the indigenous representatives and to recognize the indigenous movement as a political actor. Hence, it transformed the spatiality of politics in two ways: First, it turned public spaces like streets and plazas into political arenas where political issues and demands are deliberated. Opening up the boundaries of what constitutes a political space, it challenged the exclusionary character of more traditional spaces of politics like the national congress or the municipality. Second, by entering the national congress, the marchers challenged the white/*mestizo* racialization of the space of the national congress, which in 1990 still did not included a single indigenous representative.

Figure 5: Campaign propaganda of Pachakutik *(Source: picture author)*

1996 The emergence of Pachakutik

In 1996, the Confederation of Indigenous Nationalities of Ecuador (CONAIE) formed the indigenous political party *Pachakutik*. The name '*Pachakutik*' referred not only to the name of a legendary leader of the Andean pre-Columbian civilization of Tawantinsuyu, but it is also the indigenous word for 'a new awakening or beginning of a new era' and is thus used metaphorically (on the development of the Ecuadorian indigenous movement and the creation of Pachakutik see Andolina 1999, Becker 2008b, Collins 2001, Lluco 2005, Lucero 2008, Sánchez-Parga 2007, Selverston-Scher 2001, Yashar 1999, 2006b). The foundation of the party was an important step both for the institutionalization of indigenous politics and the indigenization of electoral politics through the bodies of indigenous politicians and their political agendas. As Becker (2008a: 184) highlights,

'Pachakutik was an explicit reversal of a policy that CONAIE adopted at its third congress in 1990 not to participate in elections because neither the political system nor political parties were functioning in a way that represented people's interests. Increasingly, however, Indigenous activists believed it was time not only for them to make their own politics but also to make *good* politics that would benefit everyone' (emphasis CS).

While the indigenous movement initially positioned itself as a social movement in opposition to institutionalized politics, with the creation of the political movement *Pachakutik*, indigenous subjects became part of electoral politics. The foundation of *Pachakutik* can be understood as the culmination of CONAIE's drive to insert indigenous peoples directly into institutionalized politics with the aim to give them a voice and allow them to speak for themselves. In its first contest in 1996, *Pachakutik* won 8 of 82 National Congress seats and 68 seats in local elections. Further, in eight provinces indigenous mayors were elected (Beck and Mijeski 2004). The presence of indigenous politicians in both national and local electoral politics challenged not only the white and *mestizo* racialization of these spaces, but also the liberal state model based on the idea of homogeneity and *mestizaje* that had rendered indigenous peoples and issues invisible during centuries (Beck and Mijeski 2000, Radcliffe 1997, Safa 2005, Stutzman 1981, Walsh 2008a). This exclusionary notion of citizenship was radically questioned by *Pachakutik's* demand to recognize the plurinationality of the Ecuadorian society within the new

Constitution of 1998. While *Pachakutik* has been successful in implementing its demands for *plurinacionalidad* and *interculturalidad*[6] in both the Constitution of 1998 and 2008 (Andolina 2003, Radcliffe 2011, Walsh 2008b, 2009), the idea of equality in difference on which these concepts are based contrasts with the political, economic and social inequalities and injustices in everyday life.

Despite the persisting discrepancy between indigenous people's everyday reality and constitutional promises, indigenous political subjects have radically challenged the notion of citizenship and political space: with regard to citizenship, the indigenous movement has introduced the concept of differentiated citizenship which is both similar and different from previous modes of interaction between indigenous people and the state. As Pallares (2007: 153–154) highlights, on the one hand, it is connected to a long tradition of indigenous autonomy and a separate indigenous republic (see also chapter 3). On the other hand, today's struggle for 'equality in difference' questions and challenges the existing universal liberal citizenship model in ways that previous struggles did not. Indigenous people are not longer protégés of patronage relations, but central participants in state formation who have been successful in placing collective demands within the framework of an embodied pluri-culturalism. The institutionalization of indigenous politics, however, has not only challenged the notion of a universal liberal citizenship, but also the very spatiality of electoral politics. Like radical democrats, *Pachakutik* rejects the

> 'sharp distinction between the public sphere of government [...] and the private sphere of family and voluntary organizations that the Western, liberal model of representative democracy delineates' (Van Cott 2008: 13).

Indigenous notions of political organization fuse both spheres through leaders who perform administrative, economic, law enforcement, and spiritual roles at the same time and through positioning the family as the basic unit of politics. In so doing, indigenous politicians have produced a new spatiality of institutionalized politics that challenges the Western private-public dichotomy defining colonial and post-colonial spaces of politics. Further, representing a social movement *and* a political party, the indigenous movement and its political arm *Pachakutik* thwart

---

[6] In order to clarify the differences between plurinationalism/pluriculturalism, *interculturalidad* and multiculturalism it is important to take into account that each of the concepts has a different genealogy. While pluriculturalism (convivality of different cultures in a post-colonial context) and multiculturalism (inclusionary occidental model of a neoliberal state) are descriptive terms that indicate the existence of different nations/cultures in a specific territory, *interculturalidad* still does not exist but is a project to be constructed. *Interculturalidad* means that there are two distinct cosmologies at work, a Western and an Indigenous one. *Interculturalidad* aims to construct a new society on the basis of a dialogue between these two epistemologies over the erosions of the liberal and republican state (Mignolo 2005, Walsh 2008a, 2009).

the emergence of clear-cut boundaries between institutionalized and non-institutionalized politics. They deliberately do so not only in order to challenge a Western liberal conception of politics, but also to position themselves on the 'good' side of social movements and distance themselves from the 'bad' (read: dirty, corrupt, clientelist) side of party politics. As the above cited quote from Becker (2008a: 184) shows, a crucial motivation to found *Pachakutik* was to make their own politics, to make *good* politics that benefits the (indigenous) people.

This claim of doing politics in a different and better way has given rise to my research interest in the everyday political practices of new political subjects. It was this promise for *good* politics that inspired me to engage with the question whether the presence and activities of new political subjects contribute to the decolonization and radical transformation of Ecuadorian politics. My research interest emerged at a time, when *Pachakutik* and the indigenous movement in general were increasingly criticized for reproducing social inequalities that prevent women and other disadvantaged groups from exercising full citizenship (Pacari 2005, Tibán 2005, Tuaza 2009, Van Cott 2008). The paradox that new ways of exclusion emerge in a setting that claims for itself to be inclusive and that indigenous women somehow did neither benefit from the rise of the indigenous movement nor from the gender quota law, aroused my interest to look at the recent political transformations in Ecuador from an intersectional perspective.

Figure 6: Leaflet of the National Council of Women (CONAMU) to foster women's participation in electoral politics (Source: CONAMU 2006)

1997 Ley de Amparo Laboral

Ecuador is not only the first country in Latin America to give women the vote, but also to introduce a gender quota law through the *Ley de Amparo Laboral* in 1997. Silvia Vega (2004: 44), a feminist academic activist, vividly recalls the introduction of the law:

'While women organizations had demanded gender equality in electoral participation since the return to democracy in 1979, the law totally surprised us; it was an initiative from the legislative block *roldista*, the government party at that time and it was introduced without consulting nor knowledge of the women organizations'.

The law was primarily designed to foster gender equality in labor relations, but women organizations appropriated the law for their ends and succeeded in applying it to the electoral law (*Ley de elecciones*). Only a few months after the law had been passed, the overthrow of President Abdalá Bucaram in February 1997 opened a window of political opportunity for women's movements. Women organizations managed to both present candidates and their political agenda within the constitutional processes that took place as a result of Bucaram's overthrow. In the elections for the constitutional assembly, the gender quota law of 20 percent was applied for the first time (Vega Ugalde 2005: 172). Since then, the quota has been raised with a 5 percent increase with each electoral cycle, reaching parity in the elections of 2009 (Cañete 2004a, 2004b, Vega Ugalde 2004, 2005). Despite the progressive legislation, women organizations like the Parliamentarian Commission of Women, the National Council of Women *(Consejo Nacional de Mujeres)* and the Political Coordinator of Women (*Coordinadora Política de Mujeres*) have fought a long struggle to secure the implementation of the gender quota law and the strict alternation between men and women on the electoral lists inscribed in the law (Cañete 2004a, Vega Ugalde 2004, 2005). The number of female candidates and politicians has significantly risen on all political levels as a result of the quota law.

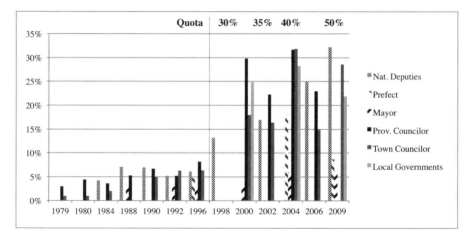

Figure 7: Women's share in political offices since 1979
Source: Own compilation based on Arboleda 1993, Brito Merizalde 1997, CONAMU 2002, Vega Ugalde 2005

The increasing presence of women in the National Congress and the local governments genders these spaces, which have been dominated by men since colonial times, in new ways (Molyneux 2000, Mosquera Andrade 2006). Hence, women have gained access to electoral politics in more significant numbers, in a similar fashion as indigenous people since the end of the 1990s. This means, that the spaces of electoral politics have been confronted with major transformations regarding the gender and ethnicity of the bodies that constitute and shape these spaces (Burbano de Lara 2004). While the spaces of Ecuadorian politics have been constituted exclusively by white/*mestizo* and masculine bodies and interests since colonial times, the increasing presence of women and indigenous people has challenged the hegemony of white/*mestizo* men within these spaces. As a result, gender, ethnicity, and class intersect in new modes within the spaces of electoral politics.

The success of both movements to institutionalize their demands and to turn themselves into recognized actors within electoral politics has similar reasons: Both movements have gained strength within the context of new social movements that appeared throughout Latin America (Alvarez, Dagnino, and Escobar 1998, Escobar and Alvarez 1992, Slater 2008) as a response to political instability and neoliberal economic reforms (Cañete 2000, Carrière 2001, Lind 2000, 2001). Both women and indigenous politicians entered electoral politics with the promise of making 'better' politics and attending to the demands of the (poor) people, in short, to right the wrongs of former governments. But there is one more commonality: In the process of political institutionalization both movements have produced further exclusions. Electoral statistics show that indigenous women benefit to a lesser extent from the quota law than *mestiza* or Afro-Ecuadorian women (see Pacari 2005, Radcliffe, Laurie, and Andolina 2002). Hence, although gender and indigenous politics have claimed to create more inclusive spaces of politics, they have produced new exclusions that (re-)produce and result from gendered and racialized political cultures (see chapter 7).

The four narratives of the introduction of suffrage, the uprising, the creation of *Pachakutik*, and the implementation of the gender quota have shown, for one thing, the close relationship between social movement and electoral politics. For another, they evidence that gender and ethnicity intersect within the electoral politics in complex ways in different historical moments and spatial contexts. Both aspects are crucial for developing a framework of a feminist electoral geography. Aiming to rethink electoral geography, this book conceptualizes the relationship between social movement and electoral politics as a central element of a feminist perspective on institutionalized politics. Further, it engages with the intersectional relations between gender, ethnicity, class, and electoral politics and the way these relations constitute and shape the discursive and material spaces of local electoral politics.

TRAVELING THROUGH ECUADORIAN LOCAL POLITICS

This work takes the reader to a voyage through the places and spaces of Ecuadorian local politics. As any journey across unknown territories, it is not a simple, direct route from A (research question) to B (research results). Rather, the journey makes loops, at some points the views from the route might be fascinating, at other points they might be boring and might seem well-known. In an attempt to give the reader better orientation along this messy road, I have decided to mark the path with road signs that provide some measure of guidance. The journey first provides a space of DEPARTURE, where preparations are undertaken for this journey. Chapter 1 (Introducing Political Change in Ecuador) positions this research within the disciplinary fields of feminist political geographies and electoral geography (Searching for a Feminist Electoral Geography) in order to provide the reader with the necessary knowledge to situate this study on her/his disciplinary and interdisciplinary imaginary map. The section 'Puzzling over Political Transformations in Ecuador' offers the reader an overview over the historical and contemporary political context of Ecuador that is at the heart of this research. Chapter 2 discusses in more detail how we can think processes of political change from a geographical perspective (Theorizing Spaces of Democracy) and suggests ways of methodologically capturing these processes (Studying Political Practices in Local Spaces of Democracy). In so doing, this first section, DEPARTURES, prepares the reader with the necessary contextual and conceptual knowledges that she/he needs to be able to follow the route of the journey through the electoral geographies of Ecuadorian politics.

Along the road the reader is offered three different theoretical lenses that shape and sharpen her/his views on Ecuadorian politics in particular ways. The first lens carries the label PERFORMATIVITIES and aims to make the reader see both the spaces and subjects of politics as outcomes of performances and performative power relations. Chapter 3 (The Performativity of Post-colonial Politics) asks how Butler's concept of performativity helps to analyze the materialization of processes of political change in political geography. Bringing Butler's notion of performativity into dialogue with Mouffe's concept of politics as antagonism/agonism, chapter 3 proposes a conceptual framework that focuses attention on the performative practices and antagonistic power relations that constitute spaces of politics. Recent political transformations in the province of Chimborazo in Ecuador serve to discuss how political change is performatively enacted through antagonistic struggles that challenge the hegemonic performative spaces of politics. Along with the Ecuadorian political imaginary of *interculturalidad*, I reflect in chapter 3 whether intercultural politics could actually be considered a first step towards an agonistic democracy. While chapter 3 sets out the conceptual framework for a performative political geography, chapter 4 (A Visual Ethnography of Political Performances) asks how such a performative perspective can be methodologically redeemed. It suggests visual ethnography as a suitable methodology for performative geographies, since it focuses explicitly on the embodied and non-textual performances that bring both subjectivities and spatialities into being. Drawing on a visual ethnographic case study of politicians' identity performances in the province of Orellana, I show how the filmed identity performances can be linked and contrasted to hegemonic discourses around masculinity, femininity, whiteness, and indigenousness represented in Ecuador's visual culture. The visual ethnography presented in chapter 4 reveals the ambivalence of their identity performances, in which the politicians are constantly torn between responding to and simultaneously resisting hegemonic discourses around the masculinity and whiteness of the political.

The second lens, labeled EMOTIONS, zooms in on a particular kind of embodied and spatialized performance. Looking at spaces of electoral campaigning through the lens of emotional geographies enables the reader to better understand the affective dimension of (Ecuadorian) politics that plays a crucial role in contemporary Latin American populism in particular. In chapter 5 (Performative Emotions in Electoral Campaigns), I develop a performative understanding of emotions that facilitates linking the emotional performances of electoral campaigns to place-specific histories of contact between different gendered, racialized and classed bodies. Conceptualizing emotions as performative, chapter 5 further develops the first theoretical lens of performativity by bringing it into dialogue with current work in emotional and non-representational geographies. The chapter compares the observed emotional performances of female and indigenous local candidates in a campaign in the Amazon province of Orellana with the emotional performances of male and *mestizo* national populists who have dominated Ecuadorian politics throughout the past decades. The comparison shows similar emotional patterns turning around a Manichean rhetoric of *rabia* (rage) and *amor* (love). It reveals, however, that the emotional performances of different candidates performatively generate the gendered, racialized and classed boundaries of *el pueblo* (the people) in different ways in particular times and places. Chapter 6 (The Intersectionality of Emotions in Campaigns) further engages with the relations between emotions, campaigns and differently marked bodies by suggesting an intersectional perspective on emotions. I depart from the assumption that emotional campaign performances both result from and respond to gendered, racialized and classed inequalities within Ecuador's post-colonial society. Chapter 6 sets out to examine the emotions new political subjects in the provinces of Orellana and Esmeraldas put on display in a local campaign, focusing on how their emotional performances respond to and challenge intersectional systems of oppression such as racism, classism and heteronormativism. The two empirical case studies in chapters 5 and 6 illustrate the claim that electoral geographies need to be more attentive to the emotional dimension of electoral spaces to understand the affective dimension of contemporary (populist) politics.

The third lens, INTERSECTIONALITY, focuses on the interdependencies between different identity categories such as gender and ethnicity and engages with the interplay between different power structures. Looking at the way access to spaces of politics in Ecuador is regulated through intersecting power relations that marginalize certain subjects on the basis of their intersecting gendered, racialized, classed etc. identity, chapter 7 engages with the power relations that structure these spaces. In a first step, an intersectional analysis of local electoral results shows that access to electoral politics – in form of successfully running for office – is uneven not only across different social groups, but also across space due to the particular (post-)colonial histories of different localities. On the basis of this finding, chapter 7 advocates grounding the concept of intersectionality in the historical and geographical context of a political locality. In a second step, the chapter discusses the political agendas of local women politicians to evaluate the impact of the quota on substantive representation. Drawing on three case study localities in Orellana, Esmeraldas, and Chimborazo, a typology is developed that presents the three most common political agendas women prioritize on the basis of their roles as mothers, teachers, or feminists. I show in chapter 7 how these political agendas are closely linked to the way the socio-material spaces of local politics are (en-)gendered. Chapter 8 engages with the concept of intersectionality from a slightly different angle by reflecting critically on how the intersectionality of our own identity enables us to build relations with our informants in the field. Engaging with the challenges of doing fieldwork as a Western researcher in the Global South after the (feminist) postcolonial turn, chapter 8 addresses the politics of fieldwork of this research project and questions how knowledge can be produced in collaborative, participative ways. Together with the co-author of this chapter, Dörte Segebart, I discuss the potentials and limits of participatory (action) research and intersectionality as two central feminist postcolonial approaches.

Theorizing political change as performative and antagonistic (chapter 3) and investigating electoral politics through an appropriate methodological approach of (visual) ethnography (chapter 4 and 5), I show how the conceptualization of political subjectivities and spatialities as performative (chapter 3, 4, and 5), emotional (chapter 5 and 6) and intersectional (chapter 6, 7 and 8) contributes to a feminist electoral geography. In the last chapter, CONCLUSIONS, I sum up the empirical results of my research and the conceptual contributions of the different chapters to the development of feminist electoral geographies. By weaving together the different conceptual approaches in this last chapter, I hope to leave the reader with a more precise, but still colorful mosaic picture of Ecuadorian local politics.

2.
RETHINKING ELECTORAL GEOGRAPHY

In the introduction, I have claimed the need to rethink electoral geography from a feminist perspective. Before I start to suggest a theoretical (Theorizing Spaces of Democracy) and methodological (Studying Political Practices in Spaces of Democracy) framework, I briefly sketch the main characteristics of the field of electoral geography. The engagement with the criticisms electoral geography has been confronted with serves as a starting point to develop what I then call feminist electoral geographies.

Electoral geography is often pictured as one of the key topics of political geography as already the founding fathers of modern geography like André Siegfried (1913) or Carl Sauer (1918) engaged with the mapping of electoral results and districts. It was not until the 1970s, however, that geographers started to deal in more systematic ways with elections (Taylor and Johnston 1979), mainly as a result of the so-called quantitative revolution of geography. Flint and Taylor (2007: 196) point out that while 'political geography was passed by in these intellectual upheavals, [...] this was not so for electoral geography'. Taylor (1978: 153) considered elections a 'positivist's dream' with a bonanza of (quantitative) data inviting researchers to engage with. Political geographers eagerly did so, mainly through applied standard statistical analysis. For the mere volume of work produced, Taylor and Flint (2000: 235) contend that 'electoral geography has been the success story of modern political geography'. Electoral geography's historical roots in the quantitative paradigm have had lasting impacts on the development of the field until today, resulting in recent critiques of its positivist framework. Despite the importance of democracy as 'the most significant global trend' (Barnett and Low 2004: 1) and the 'analytical potential afforded by electoral spacings' (Cupples 2009: 111), electoral geography has been largely dominated by a-theoretical, quantitative and positivist approaches like Taylor and Flint (2007), Barnett and Low (2004) and Johnston and Pattie (2004, 2006), and others have indicated. As Agnew points out, Richard Morrill (1981, 1987), Ron Johnston and Charles Pattie (2004, 2006, 2008, 2013), Colin Flint and Peter Taylor (2007), and John Agnew (1996) himself (note it is a 'men only' club!) have since attempted to revitalize electoral geography by integrating geographical and spatial theory into electoral analysis.

In short, these studies have been concerned with the geographies of representation and the effect of place on political behavior. Despite these attempts, Agnew (1990: 20) criticizes that because of the domination of positivist approaches in electoral geography, the sub-discipline still believes that the empirical findings 'speak for themselves' and engages with methodological rather than theoretical discussions. While Agnew's critique refers to work published between 1960 and

1987, Taylor and Flint (2000: 236) more recently have taken up his critique suggesting that electoral geography has no clear intellectual strategy beyond the conduct of empirical inquiry and therefore is not contributing to a coherent body of knowledge which can inform wider debates regarding democracy and its implementation. In 2004, Johnston and Pattie (2004: 46) conclude their review of the current state of the art in electoral geography with the judgment that 'Agnew's criticism still carries much weight'. They state that 'the field has changed insufficiently for an entirely new framework to be introduced at this stage'. I suggest, however, that it *is* time for a new theoretical and methodological framework in electoral geography. I develop this new framework on the ground of a number of recent studies that have productively integrated poststructuralist thinking into electoral geography.

As the feminist and cultural turn within political geography has fruitfully broadened what counts as political subject matter (see Staeheli and Kofman 2004a: 6), poststructuralist thinking has first been employed in the engagement with these new political spheres and topics such as social movements, the body, the home etc. (e.g. Dowler 1998, Marston 1990, McEwan 2000). While the refocusing on these 'small p' political spaces (in contrast to the 'capital P' spaces of state politics) is in general a positive move, institutionalized politics such as elections and political parties have rather been neglected by poststructuralist scholars (Barnett and Low 2004: 7). This trend has to be understood more as a corrective to the narrow, masculinist, state-centered view of mainstream political geography and geopolitics (see Dowler and Sharp 2001, England 2003, Hyndman 2001) than as feminists' lacking interest in institutionalized politics. While feminist political geographers widely recognize the importance of state institutions for social change, only few have engaged empirically with electoral politics. This work, especially of feminist and queer geographers, however, is considered crucial for the rethinking of electoral geography:

> 'Some of the most recognizable examples of changing epistemologies manifested themselves in the studies of identity and electoral politics. Race, gender, and sexuality were all brought into focus as contestations of spatialized identities as played out within electoral processes' (Leib and Quinton 2011: 19).

In what remains of this section, I discuss some of this work with the aim to identify first traces of an emerging feminist electoral geography in the discipline and by doing so acknowledge the work that has inspired my own approach.

Anna Secor's empirical work on Islamist politics in Istanbul (2001a, 2001b) and her conceptual contribution 'Feminizing Electoral Geography' in 'Mapping Women, Making Politics' (2004) are an inspiring starting point to think through electoral geography from a feminist perspective. In the latter, Secor engages with the geography of women's representation in political office and the role of gender as a political cleavage. Focusing on findings from the Middle East, the UK, and the US, she is interested in the way 'women's participation is contextually embedded and contingent on different conceptions and performances of gender relations' (Secor 2004: 262). Secor's contribution to the development of a feminist electoral geography is twofold: On the one hand, she focuses on women as political subjects within electoral politics. On the other hand, by engaging with women in Turkish politics, she challenges the Western bias inherent in feminist electoral studies.

Staeheli's (2004) survey based approach to the question whether difference makes a difference in political participation in both informal and electoral politics is very informative for thinking gender and difference together. Departing from the premises of a politics of difference, Staeheli analyses political activism of men and women in four US cities. On basis of her findings, she suggests that it is crucial to ground women's political actions in more mundane politics and activities of urban life. For her, 'such grounding is important so as to avoid overstating the implications of women's activism and inadvertently essentializing it' (Staeheli 2004: 365). Her call to be cautious about any romanticizing and essentializing tendencies with regard to women's political participation (both in electoral and informal political spheres) is crucial for the development of a feminist electoral geography.

While Secor and Staeheli take gender as primary category of analysis, Brown, Knopp, and Morrill (2005), Nelson (2006) and Radcliffe, Andolina, and Laurie (2002) engage more explicitly with the way gender intersects with other identity categories relevant in the specific contexts researched. In their article 'The culture wars and urban electoral politics: sexuality, race, and class in Tacoma, Washington', Brown et al. (2005) explore how these three categories are co-constructed in part through the discourses and practices of elections. Drawing on a qualitative analysis of the campaign rhetoric and a quantitative analysis of the election's socio-demographics, they look at the cultural dimensions of gay rights issues within particular demographic and electoral configurations. Their approach is inspiring as they think race, class and sexuality in an intersectional way both in their quantitative and qualitative analysis.

In her account 'Geographies of state power, protest and women's political identity formation in Michocacán, Mexico', Lise Nelson (2006) engages with the question how race and gender as nonessential categories intersect differently through space. Comparing indigenous women's narratives of protest after the massive electoral protest of 1988–1989 in three indigenous communities of Michoacán, Mexico, she shows that the differences that emerged between women's political identities and discourses in these three communities were directly shaped by the community's 'geopolitical positioning' within broader configurations of nation, race, and state territoriality in Mexico. Her emphasis on the necessity of linking local political practices to the broader post-colonial geopolitical processes is essential for a feminist electoral geography, which seeks to challenge the narrow focus on the level of national politics dominant in conventional electoral geography. Further, her engagement with political actions in a post-colonial context like Mexico contributes to a feminist electoral geography sensitive to contexts other than Western democracies that have been at the center of electoral geography over the last decades. In engaging with post-colonial democracies, a feminist electoral geography aims to open up discussions about alternative understandings of democracy in order to challenge any universal meaning of democracy. Engaging with the specific meaning of democracy in different time-spatial contexts, feminist electoral geographies need to be attuned to the spaces in which politics takes place and the spatialities of political performances (this call was also articulated by Flint and Taylor 2007).

In the paper 'Reterritorialised space and ethnic political participation: Indigenous municipalities in Ecuador', Radcliffe, Andolina and Laurie (2002: 289) show 'how the inclusionary impulse of decentralization legislation and social movement politics is limited by persistent racial and gendered political cultures'. Their work contributes to a feminist electoral geography by focusing on the inclusions and exclusions produced in political processes of transformation. Especially Radcliffe's later work draws more explicitly on theories of intersectionality to explore the way gender, ethnicity, race, class, and locality shape the opportunities of women to participate in institutionalized and non-institutionalized politics (Radcliffe 2010, Radcliffe and Pequeño 2010). While Radcliffe et al.'s research has focused mainly on the political participation of indigenous women in terms of electoral politics and social movement politics, building on their work, my research is interested to widen their focus by engaging with women with different ethnic backgrounds in local electoral politics.

While there is more theoretically inspiring work dealing with issues of elections, political parties and institutionalized politics (e.g. Bélanger, Carty, and Eagles 2003, Cupples 2009, Schaffner 2006, Warf and Leib 2011), the studies presented here have been selected on the ground that they contribute explicitly to the development of feminist and/or postcolonial electoral geographies. Table 1 recapitulates the main points of the presented literature in comparison to more conventional electoral geography. Building on the studies discussed and my own empirical work on local electoral politics in Ecuador, I suggest in the following a more comprehensive theoretical and methodological framework for a research agenda of feminist electoral geographies.

Comparing the research agenda of 'conventional' and 'new' electoral geographies		
	Conventional electoral geography	**New electoral geography**
Epistemological grounding	▪ positivist paradigm	▪ cultural, feminist, poststructuralist turn
Theoretical background	▪ a-theoretical, positivist ▪ 'place perspective'	▪ theories of difference, ▪ intersectionality ▪ performance studies ▪ post-structuralism
Methodology	▪ Quantitative ▪ standard statistical analysis	▪ multi-method ▪ quantitative ▪ qualitative ▪ ethnographic ▪ collaborative

Empirical foci	<ul><li>spatial variations in voting patterns</li><li>geographical influences on voting</li><li>geography of representation</li><li>geography of electoral systems</li></ul>	<ul><li>construction of difference within and across political spaces</li><li>intersections of gender, race, ethnicity, class, sexuality in electoral politics</li><li>intersectional geographies of inclusion and exclusion in electoral participation and representation</li><li>time-spatial specificities of democracy</li></ul>
Geogr. contexts	<ul><li>Western liberal democracies</li></ul>	<ul><li>democracies across the world</li></ul>

Table 1: Comparing the research agenda of conventional and new electoral geographies
Source: Compilation of the author

THEORISING SPACES OF DEMOCRACY

In this section, I present the theoretical approaches I have employed in my work – namely antagonism, performativity and intersectionality – and show how each of them has helped me to understand the political transformation processes taking place in Ecuadorian local politics. At the same time, I suggest these three approaches as suitable theoretical foundations for a feminist electoral geography. Discussing each theoretical approach, I show how each contributes to opening up the positivist and empiricist framework of electoral geography (Flint and Taylor 2007: 196, Johnston and Pattie 2004: 46) and developing a feminist framework for electoral geography.

Understanding (Spaces of) Politics as Antagonistic

Staeheli and Kofman (2004a) have identified three different (but related) approaches in feminist political geographers' understandings of the 'political'. They differentiate between approaches in feminist political geography that conceptualize the political as antagonism, as constitutive, and as distribution (see also Brown and Staeheli 2003). *Antagonistic* approaches depart from the assumption that the conflictual dimensions of any interaction characterize the political. In so doing, they draw attention to the processes of interest formation, coalitions, and place making that shape political struggle. Feminist political geographers have shown how in different political contexts antagonistic relations between a constructed (and often gendered and sexualized) 'us' versus 'them' characterizes a whole range of political articulations from (violent) political conflicts (Dowler 1998) to identity politics (Knopp and Brown 2003). The *constitutive* implies understanding the political as an ongoing process in which societies and 'the political' are made in and through social struggles. Research in this strand highlights the contingencies of political struggles and emphasizes that 'politics in this sense is constitut*ive*, rather than constitut*ed*' (Brown and Staeheli 2003: 253). *Distributional* approaches in feminist political geography engage with the distribution of power, resources, and privileges, asking who gets what under which circumstances. Seagar's (1997) 'The State of Women in the World Atlas' is an example of how feminist geographers analyze spatial patterns of inequality.

I would like to adapt their systematization to post-colonial contexts by positioning antagonism at the center of both the 'political' and 'politics' in post-colonial societies. Placing antagonism at the center of the 'political' and 'politics', I argue that antagonism defines further the constitutive and distributional dimension of the 'political' and 'politics' in post-colonial settings. I develop my approach to feminist electoral geography by departing from the assumption that the 'political' and 'politics' are inherently antagonistic. Then, I discuss the differentiation between the 'political' and 'politics' that has been central in debates in feminist political geography. Finally, I propose a feminist perspective to think through the spatialities within which democratic politics takes place.

Slater (2002: 257) points out that for the societies of Latin America, 'the tenets governing the constitution of their mode of political being were deeply molded by the fact of conquest'. He argues that the 'framing of time and the demarcation and ordering of space followed an externally imposed logic' (ibid.). The demarcation and ordering of space played an important role in the (re-)production of colonial power. Goss (1996) shows that public spaces like the plaza, which are considered as important political venues until today, are colonial impositions (on the colonial history of the spaces of electoral politics see also chapter 3). The construction of public spaces aimed to re-inscribe the Western public-private separation in the colonial landscape. By so doing, the public spaces (re-)produced hierarchies by excluding certain bodies from participating in civic events taking place in these spaces. Hence, the regulation of access to public spaces produced social hierarchies in colonial societies (Blunt and Rose 1994, Mills 1996, Wilson 1997). The effects of the colonial ordering of (public) space still reverberate in the post-colonial period. The contestation of this colonial ordering through acts of resistance and movement protests forms an essential part of post-independence politics. Hence, the antagonistic 'struggle over inclusion [between (post-)colonial powers and colonially marginalized groups] is central to building and consolidating democracy in postcolonial settings' (Staeheli and Mitchell 2004: 155). Feminist political geographers have shown that the definition of what counts as political and where the political is located (public = political, private = a-political) can itself be considered an outcome of these struggles (Cope 2004, Sharp 2003, Staeheli 1996, Staeheli and Mitchell 2004). Due to the colonial history of post-colonial democracies that have defined and placed the political on the ground of colonial power relations, antagonistic struggles over the hegemonic order are a constitutive element of the political. Hence, within post-colonial societies antagonism plays a particular role in the negotiation of a political order that differs from antagonisms found in Western democracies.

I suggest Mouffe's conceptualization of the political as antagonistic and her agonistic model of democracy as a productive way to engage with the antagonism in post-colonial societies (Laclau and Mouffe 1985, Mouffe 1995, 2005a, 2005b, 2008)[7]. To understand her approach, her distinction between 'the political' and 'politics' is crucial:

> 'By "the political" I mean the dimension of antagonism which I take to be constitutive of human societies, while by "politics" I mean the set of practices and institutions through which an order is created, organizing human coexistence in the context of conflictuality provided by the political' (Mouffe 2005b: 9).

The distinction between 'the political' and 'politics' as outlined by Mouffe has inspired the development of theories of poststructuralist radical democracy as they situate the contestation over the boundaries of 'the political' at the center of democratic politics. The institutions and practices of democratic politics like the executive, legislative, and judiciary institutions that constitute democracies and the practices taking place within these institutions, such as designing laws and political actions that aim to create order, can be considered as outcomes of antagonistic struggles over how to create order. As the order that is created through the practices and institutions of politics has a hegemonic nature, this order is itself the product of antagonistic struggles and subject to constant contestation.

Mouffe's (2005b: 10) understanding of the 'political' as antagonism needs to be considered as a response to liberal thought in political philosophy that is characterized by a rationalist and individualist approach which forecloses acknowledging the nature of collective identities. Mouffe criticizes liberalism for its rational view of the political as a 'harmonious and non-conflictual ensemble' that negates the antagonistic dimension of the 'political'. Drawing on Carl Schmitt's friend/enemy discrimination as criteria of the political, she understands the conflictual nature of politics as starting point for democratic politics. In Mouffe's (2005a: 103) view, the aim of democratic politics is 'to transform antagonism into agonism' by providing channels through which collective passions will be given ways to express themselves (on the role of emotions in electoral politics see chapter 5 and 6). Hence, the prime task of democratic politics is not to eliminate passions from the sphere of the public, in order to render a rational consensus possible, but to mobilize those passions towards democratic designs (ibid.). What de-

---

7 Mouffe's work is often closely associated with the work of Ernesto Laclau as they co-authored 'Hegemony and Socialist Strategy' (1985) where they redefine Left politics in terms of radical democracy. While I do draw on this earlier work of Mouffe that she elaborated jointly with Ernesto Laclau, I concentrate on her later – single authored – work that engages more profoundly with the question how an agonistic pluralism can be established as political order. For an extensive account on Mouffe's thinking and how it can be employed to think about political change in post-colonial contexts see chapter 3 and Schurr (in press).

mocracy requires is drawing the we/they distinction in a way which is compatible with the recognition of the pluralism constitutive of modern democracy (Mouffe 2005b: 14).

Mouffe's emphasis on the importance of pluralism as a constitutive element of democracy makes her theoretical approach especially fruitful for post-colonial societies in which liberal models of democracy are increasingly contested by social movements that demand the recognition of difference and pluralism (Fraser 2005, Fraser and Honneth 2003, Staeheli 2008). Defining the political as a realm of conflict, contestation and antagonism, these struggles turn into the center of post-colonial politics. In post-colonial societies, post-colonial institutions of politics are challenged to transform the antagonism resulting from colonial and post-colonial power relations into a pluralist democracy based on agonism – the struggle between adversaries (Mouffe 2005a: 102) (for an example of what such a politics of agonism could look like see chapter 3). In Mouffe's definition, 'politics' is closely linked to 'the political' as 'the political' is a constitutive element of 'politics'. Conceptualizing 'the political'/'politics' in such a co-constitutive way, any dichotomization between 'the political' and 'politics' is challenged. Thus, the concept of radical democracy as developed by Mouffe expands the boundaries of politics by 'involving participation in a range of formal and informal practices of identification and opinion formation' (Barnett 2004: 8). At the same time, it refocuses attention on what is ordinarily defined as 'politics' – with matters of policy, legislation, elections, and parties – issues that are often dismissed in more radical accounts of counter-hegemonic politics.

I now argue that Mouffe's radical democracy offers the possibility to refocus attention on formal political institutions of the state within feminist political geography by defining 'the political' and 'politics' in a co-constitutive way. Feminist political geographers have long criticized the sub-disciplines of political geography, electoral geography and geopolitics for centering on the sphere of formal (inter-)national politics and neglecting the political activities (of women, in particular) taking place in private, informal spaces and spheres (Dixon and Marston 2011, Dowler and Sharp 2001, England 2003, Hyndman 2001, 2004, Kofman 2008, Kofman and Peake 1990, Pain 2009, Sharp 2004b, Staeheli, Kofman, and Peake 2004). Through rich empirical work, feminist political geographers have demonstrated that the private sphere is an important site in which political subjects, ideas, and strategies are produced (Brown 1997, Cope 2004, Fincher 2004b, Fluri 2011b, Marston 1990, McEwan 2005).

While their intervention has been and still is very important to open up the narrow masculinist, state-centered, big 'P' understanding of politics in political geography, I argue that simultaneously we need to refocus on institutionalized politics as people's lives are still shaped by the actions that are taken (or not) in institutionalized politics. In order to understand the changing dynamics of formal political institutions of the state, we need to look closer at the complex relationship between 'the political' – understood in Mouffe's sense as the antagonistic relations in a society – and the institutions of the state – considering electoral governments at all levels an important site of institutionalized politics. Following

Mouffe's distinction of 'the political' and 'politics', I argue that a feminist electoral geography needs to focus on the interplay between electoral politics and the political action of non-institutionalized political actors. Paying attention to the antagonistic relations that constitute a society, a feminist electoral geography is able to question the hegemonic nature as well as the in- and exclusions of institutionalized politics. In so doing, the stark opposition found in feminist political geography 'between representative forms of democratic politics, presumed to be the source of dissatisfaction, and idealized models of alternative politics' (Barnett and Low 2004: 7) can be overcome by taking seriously the interrelation between electoral politics and the antagonistic struggles of (civil) society.

The close relationship between political actions taking place in the formal political institutions of the state and in the informal public and private spaces is crucial when thinking through how 'spaces of democracy' in more broader terms and 'spaces of electoral politics' in particular can be conceptualized from a feminist perspective. Staeheli and Mitchell's (2004: 147) discussion about the locations of politics is essential for the endeavor of opening up the spatial boundaries of electoral politics which have often been pictured as restricted to the spaces of the state:

> 'The spaces of democracy are necessarily plural. They include, most obviously, the spaces constructed in and through the state, such as the nation-state, cities, governments, and public spaces. But we should also consider the possibilities for democracy in other spaces that may not seem so obviously political or so obviously constructed and maintained by the state. These spaces may be associated with the home, neighbourhood, commerce, work, worship, or any gathering place. We argue [...] that these are not just spaces of democracy or of politics; they are spaces defined by unequal and differential access'.

Staeheli and Mitchell make three points in their definition of spaces of democracy that are all crucial for developing what I call a feminist electoral geography. First, they emphasize the plurality of possible spaces of/for politics beyond the spaces of the state officially assigned to politics. Second, in their list of possible spaces of politics they explicitly include spaces that have traditionally been thought of as private and hence a-political such as the home. Third, they point out that spaces where political issues are deliberated are characterized by unequal access. What implications does this understanding of spaces of democracy have for conceptualizing a feminist electoral geography? I argue that a feminist electoral geography is challenged to localize and identify both the private and public spaces where electoral decisions are made, electoral opinions are shaped, electoral strategies are discussed, canvassing takes place, and finally where elected politicians realize their political actions once they are elected. By studying the practices, performances, and interactions of elected politicians, we can understand how spaces of politics are brought into being both outside and inside the traditional spaces of electoral politics such as the buildings of the national, regional and local governments. Departing from the assumption that the spaces of politics are created through antagonistic struggles, a feminist electoral geography recognizes that these spaces are objects of ongoing contestation, as these struggles do not necessarily result in more inclusive and just polities. Feminist research can contribute to social change by understanding through which discourses and practices the access to electoral politics is gendered, racialized, sexualized, and classed. Table 2 shows how an antagonistic approach can contribute to the development of feminist electoral geographies.

Step 1 on the way towards feminist electoral geographies			
Theoretical lens	Conceptual contribution	Empirical foci	Methodological implications
Politics/the political as antagonism (Mouffe 2005a,b)	Theorizing power relations in electoral/ institutionalized politics	Antagonisms that saturate the spaces of democracy	Sensitivity to power relations that constitute the researched context
	Overcome public/private binaries that dominate conventional electoral geography	Relationship between social movements and institutionalized politics	Need to empirically engage with both actors of institutionalized and non-institutionalized politics
	Open up boundaries of institutionalized/electoral politics beyond the spaces of the state	Search for new ways of doing politics and new spaces of democracy	Open up and decolonize Eurocentric imaginaries of democratic politics through extensive engagement with political actions in the 'field'

Table 2: Approaching feminist electoral geographies through antagonism
Source: Own compilation

Finally, I would like to argue that feminist electoral geographies need to be attentive both to the *discursive* and *material* struggles that constitute spaces of politics. While focusing on discourses (and their sedimentation in laws and practices) is crucial to understand how the exclusion of certain bodies is discursively legitimized and naturalized (see chapter 3), tracing the material construction of spaces of politics offers insight into how political agendas and practices shape the very spatiality of politics. Butler's notion of performativity that links discourse, practice and materiality seems an appropriate theoretical lens for this endeavor. In the next section, I introduce the concept of performativity and show how it can serve to conceptualize processes of political change within electoral politics (for a discussion how to think antagonism and performativity together see chapter 3).

From the Performativity of Gender to Performative Spaces of Politics

Despite the success of the concept of performativity in feminist, social, cultural and economic geography (for an overview see Dirksmeier 2009, Pratt 2000b), political geographers have scarcely engaged with the it (Rose-Redwood 2008). While I extensively discuss in chapter 3 why the concept is fruitful to capture political change, at this point, I restrict the theoretical discussion to the conceptualization of (political) identities and spaces (of politics) as performative. The aim of this section is to show how a performative approach contributes to a more vivid electoral geography by focusing on the embodied (chapter 4) and emotional (chapter 5 and 6) performances and everyday practices (chapter 3 and 7) that constitute the political subjectivities and spatialities of electoral politics. Hence, following the systematization of Staeheli and Kofman (2004), this section is interested in the constitutive element of politics.

I go back to Butler's early discussion on the performativity of gender for two reasons: First, by so doing, I conceptualize my central analytical categories gender and race/ethnicity. Second, Butler's notion of performativity serves as a starting point for the endeavor of 'Taking Butler elsewhere' – namely to geography – along with Gregson and Rose (2000).

In 'Gender Trouble', Butler (1999 [1990]: 34) states:

> '*gender* is not a noun, but neither is it a set of free-floating attributes, for we have seen that the substantive effect of gender is performatively produced and compelled by the regulatory practices of gender coherence. Hence, [...] gender proves to be performative – that is, constituting the identity it is purported to be. In this sense, gender is always a doing, though not a doing by a subject who might be said to pre-exist the deed'.

A performative understanding of identity links linguistic and embodied performances of identity to the citational practices that produce and subvert discourses. At the same time, these same discourses enable and discipline the very performances of the subjects. In short, as Gregson and Rose (2000: 441) nicely put it, 'performativity involves the saturation of performances and performers with power, with particular subject positions'. Nash (2000: 654) points out that Butler's concept of performativity can be considered an attempt to rethink the relationships between determining social structures and personal agency by suggesting that 'women and men learn to perform the sedimented forms of gendered social practices that become so routinized as to appear natural'. As a consequence, gender does not exist outside its 'doing', but its performances need to be understood as a reiteration of previous 'doings' that have been naturalized as gender norms.

For Chambers and Carver (2008: 33), the concept of gender performativity needs to be seen as a '*political battle* against naturalizing fictions and launches a practical war of the categories'. This reading of Butler's 'Gender Trouble' becomes especially powerful when linking it to the reality of (post-)colonial societies. Feminist postcolonial scholars demonstrate that colonial oppression and exploitation have taken place through the intervention of race and the re-production of a patriarchal gender order (Dietrich 2000, Lewis and Mills 2003, McClintock

1995). Colonial discourses and practices have naturalized racial and gender differences. The iterative power of these naturalizing discourses becomes visible in the effects they have until today in (re-)producing social, economic and political hierarchies along gendered and racialized differences. Mahtani (2002) has developed a rich account that explores the performativity of race in analogy to the performativity of gender as discussed by Butler. For her, 'racialized productions, like gendered productions, are culturally constructed, rather than biological, imperatives' (Mahtani 2002: 428). At the same time, she highlights that race is not a cultural construction in the same way as gender, as the former (race) is a 'science fiction' whereas sex (of which gender is derivative) is not (see also Alcoff 2006, Butler 1993, Pateman and Mills 2007, Stolcke 1993). Without entering in greater detail into the debate about the similarities and differences between the categories of gender and race, I argue with Gregson and Rose (2000) that performativity provides an important conceptual tool for critical geography to denaturalize taken for granted social categories and practices. By tracing the discourses and practices whose citational iterations have resulted in the apparent naturalization of social categories like gender or race, one can respond to Kobayashi's (1994) claim to 'un-naturalize' these discourses that reproduce spatially inscribed gendered and racialized power relations.

Feminist scholars working on issues of identity in the Andes, have productively employed Butler's notion of performativity to de-naturalize post-colonial social hierarchies. Howard (2009: 18), for example, discusses how 'in using words such as *"indio"*, *"indigena"*, *"mestizo"*, *"cholo"*, *"chola"*, *"negro"*, *"blanco"*, *"mulato"*, *"criollo"*, and others – both in our spoken interactions with Andean people and in our written analytical accounts – we "interpellate" (Althusser 1971) the subjects we seek to describe', thereby '[hailing them] into place as the social subjects of particular discourses' (Hall 1996b: 5). Valdivia (2009: 538), engaging with indigeneity in the Ecuadorian Amazon, shows how 'performativity and materialization produce the effect of the indigenous subject' through repeated acts that signify 'Indianness' such as the consumption of particular foods, clothing, verbal and body language. Radcliffe (1997) has demonstrated in her work how 'Indians' and 'indigenous-ness' have been shaped by discourses within the remit of defining and forging the nationally imagined community (in her work, however, she does not explicitly use Butler's language of performativity). In sum, Howard, Valdivia and Radcliffe show how indigeneity is performed, brought into being and shaped through linguistic interpellations, signifying practices and discursive framings. At the same time, however, they emphasize that these productions of 'indigenous-ness' are not uncontested. The possibility of disruption in the reiteration outlined by Butler (1990a, 1990b, 1995) as well as Hall's (1996b) view of 'identification' as a two-way articulation between interpellation on the one hand and active 'production of subjectivities' on the other, opens the way to resistance. Howard (2009, 2010) examines how the spread of the term 'decolonization' in Andean political discourse since the indigenous uprisings in the 1990s constitutes the very processes of change of which this shift in the use of language is part. Focusing on bodily practices, Valdivia (2005, 2009) looks at the way embodied practices (such as clothing, eating, walking, talking) are performed slightly differently in an attempt to disrupt the norms that define 'indigenous-ness'. Finally, Radcliffe et al. (Radcliffe 1997, 1999, Radcliffe and Laurie 2006, Radcliffe and Westwood 1996) have shown how the 'nature' of 'indigenous-ness' has changed over time as the state project of dealing with indigenous groups has changed.

Building on this body of literature, I argue in my work that by focusing on the linguistic and embodied practices of identity performances of political subjects – namely electoral candidates or politicians – the power relations and discursive imaginaries that saturate the spaces of politics can productively be grasped (see chapter 4). Researching the identity performances on diverse political and electoral stages, a feminist electoral geography can engage with the question of how political subjectivities are (re-)produced as gendered and ethnicized subjects in particular ways in both the performances of the politicians and the imaginations of the audience. With regard to processes of political change, a performative research agenda is able to engage with the way stereotypes of political subjects (the male, rational, white politician) are (re-)produced and contested in the everyday performances of politicians (see chapter 4, 5). Engaging with identity-specific stereotypes of political subjectivity further contributes to understand why certain (bodily marked) subjectivities have access to the spaces of politics while it is denied to others (see chapter 3, 7). By focusing on identity performances in the spaces of politics, it can be asked how political practices (re-)produce or challenge the social (gendered and racialized) norms of a particular society.

A performative approach, however, is not only productive with regard to political subjectivities, but also when engaging with the very spatialities of institutionalized politics (Schurr and Wintzer 2012). Gregson and Rose's (2000) seminal paper 'Taking Butler elsewhere' is a promising starting point for this endeavor. They integrate Butler's notion of performativity into geography arguing that not only subjectivities, but spaces too need to be thought of as produced by power. As the performances that bring spatialities into being are in fact articulations of power, spaces are always saturated with power. In my work, I expand Gregson and Rose's notion of performative space in two ways:

First, I take their performative approach to political geography, arguing that by positioning power at the center of their understanding of space, their work is especially productive for the conceptualization of spaces of politics as antagonistic (see previous section and chapter 3). A performative approach engages with the citational practices that bring these spaces into being, including both written documents like the Constitution or laws as well as everyday practices like the passing of laws, holding of speeches etc. It does so by questioning how certain routines (like the minutes of a political meeting), practices (such as passing law), imaginations (of where institutional politics is located or what a politician looks like), and exclusions (by defining some bodies as more appropriate for political representatives) become naturalized and hence hegemonic through the constant repetition and citation of certain power-saturated practices and imaginations. Analyzing the genealogies of the norms, practices and imaginations that constitute the spaces of electoral politics is a first step for a feminist electoral geography that aims to transform electoral politics towards a more participative endeavor.

Second, I argue that Gregson and Rose's notion of performative space focuses mainly on a discursive and relational understanding of space and neglects the material effects of power-saturated performances on space. Engaging with the way everyday performances shape and transform the material construction of spaces of politics, I show how their performative approach can be expanded towards the very materiality of space (see chapter 3). I argue that political transformation processes – resulting of slightly changing citations – often leave traces in material space. The integration of a women's office within the building of the municipality or the temporal relocation of the mayor's office to the communal house of a remote rural community to receive citizens for an office hour are examples of how material spatial structures render political change visible. I argue that a feminist electoral geography benefits from focusing on the material traces of political change, as they make visible even very small changes in the constitution and construction of the spaces of politics. Further, by so doing, new spaces of democracy located outside the traditional spaces of institutionalized politics can be identified and their importance acknowledged.

Step 2 on the way towards feminist electoral geographies			
Theoretical lens	Conceptual contribution	Empirical foci	Methodological implications
Performativity of identity and space (Butler 1990, Gregson/Rose 2000)	New way of theorizing identity that goes beyond the fixed and static use of gender and ethnicity as mere statistical variables in conventional electoral geography.	Construction (and contestation) of gendered and racialized identities and norms within institutionalized politics	Take into account both linguistic and embodied practices of political subjects' identity performances through interviews, observation and visual methods.
	Trace genealogies of the spaces of politics to analyze power relations of these spaces.	Discursive and material construction of spaces of politics	Identify routines, emotions, imaginaries that constitute spaces of politics through ethnography, media and document analysis.
	Grasp political change beyond electoral results by engaging with the way political subjects transform the spatialities of politics through their practices.	Materialization of political change within the spaces of politics	Challenge of comparing the before and now of political change either through interviews, oral history or long term ethnographic fieldwork.

Table 3: Approaching feminist electoral geographies through performativity
Source: Own compilation

The Emotional Spaces of Passionate Politics

In 'Passionate Politics', Goodwin, Jasper, and Polletta (2001: 1) lament that

> 'once at the center of the study of politics, emotions have led a shadow existence for the last three decades, with no place in the rationalistic, structural, and organizational models that dominate academic political analysis'.

Like political science, geography has long 'had trouble expressing feelings' (Bondi et al. 2005: 1). The 'recent emotional turn' (Smith et al. 2009b: 4), however, has forced geography to reflect 'on the extent to which the human world is constructed and lived through the emotions' (Anderson and Smith 2001: 7). So just as feminist geographies have critiqued geographical accounts of a world where women had been noticeable largely by their absence, 'emotional geographies too might be thought of as a recognition and response to a lack, that is, geography's failure to represent our emotional lives' (Smith et al. 2009b: 4). Emotional geographies are concerned with 'the spatiality and temporality of emotions, with the way these coalesce around and within certain places' (Bondi et al. 2005: 3). They understand emotion 'in terms of its socio-*spatial* mediation and articulation rather than as entirely interiorized subjective mental states' (Bondi et al. 2005: 3) as psychological accounts of emotions do. The body of research under the label 'emotiona/affectual geography' has grown rapidly over the last decade in form of edited collections (Davidson, Bondi, and Smith 2005, Smith et al. 2009c, Thrift 2008), handbook articles (Ettlinger 2010, Thien 2011, Woodward 2011, Woodward and Lea 2010) and multitudinous journal articles. The foundation of the journal 'Emotions, Space and Society' in 2008 and the already fourth International Conference on Emotional Geographies in 2013 bear witness to the success and importance of the emotional turn in geography. So far, however, emotional geographies have been oriented 'more towards social, cultural and environmental dimensions, than mapping out the political geographies of emotion' (Pain et al. 2010: 973).

The different subfields of political geography have engaged to different extents with issues of emotion. Geographies of social movements were among the first to engage with the role emotions play for political spaces. Bosco's (2004, 2006) research, for example, on the gendered emotionality of the '*madres de Plaza de Mayo*' who protest against human rights violations in Argentina builds on a wider range of literature around emotions in social movements (Goodwin, Jasper, and Polletta 2001). This work addresses collective feelings and the relation between emotions, political activism and identity such as ethnic groups (West-Newman 2004), feminist movements (Holmes 2004, Taylor and Rupp 2002a), gay movements (Gould 2004), and motherhood (Bosco 2004, 2006). A recent special issue in ' Emotions, Space and Society' has further opened up discussions about the emotional spaces of political activism (see editorial by Brown and Pickerill 2009). Emotions are in these accounts related to social movement protests, suggesting that emotions are rather experienced within informal political spaces of civil society than in electoral political spaces.

Recently, feminist geopolitics have made important inroads in studying the emotional dimension of geopolitical events and processes (Fluri 2011b, Koopman 2011, Pain 2009, 2010, Sharp 2011, Smith 2011). While feminist political geographers have long questioned prevalent dichotomies about formal/informal politics, public/private spaces and the characterization of these spaces as rational/emotional and masculine/feminine (Blunt and Rose 1994, Staeheli 1996, Staeheli, Kofman, and Peake 2004, Staeheli and Mitchell 2004), they have not yet paid attention to the emotional dimension of institutionalized politics. One reason for this can be found in feminist political geographers' endeavor to open up the narrow, masculinist, and state-centered perspective of mainstream political geography (see e.g. Dowler and Sharp 2001, Hyndman 2004, Staeheli, Kofman, and Peake 2004). In their attempt to refocus attention on diverse political settings beyond the state, such as social movement politics (Conway 2008, Gruszczynska 2009) or politics of care (England 2003, Pratt 2004), institutionalized politics and related issues such as elections and campaigning have remained on the margins of feminist political geography.

At the same time, in electoral geography 'atheoretical, quantitative and positivist approaches' (Cupples 2009: 111) dominate and have increasingly become the subject of critique (Agnew 1990, Johnston and Pattie 2004, Warf and Leib 2011). While there have been important contributions to electoral geography from a feminist perspective (e.g. Nelson 2006, Secor 2004, Staeheli 2004), the role of emotions in the constitution of electoral spaces has not been taken into account in this field. Following a feminist political agenda that calls for the need to embody, locate and ground political action (Dowler and Sharp 2001, Hyndman 2001, 2004), I develop an emotional electoral geography in this book that asks how emotions are expressed and evoked in spaces of politics. I thus call for refocusing attention on institutionalized politics as one of many sites of political struggles where political communities are generated through emotional performances.

Geographies of affect have called attention to the way the systematic engineering of affect can also be interpreted as key political technology (Thrift 2004). Thrift argues that if affect is 'intimately connected with the political and the exercise of politics', then in turn 'generating affect relies on manipulating space' (Thrift 2006: 557). Thrift's claim for the importance of understanding politics of affect in order to grasp their manipulative force certainly also gains importance in contexts like Ecuador due to the ongoing medialization and the importance of populist practices (de la Torre 2000). I argue, however, that emotions not only represent a 'threat' to the quality of democracy as Thrift frames it, but constitute also a productive power for change within the political. Drawing on Barnett's (2008: 198) critique on Thrift that feelings 'have no *a priori* political valence at all', it is important to investigate how certain feelings become politicized by whom and with which aims. Tolia-Kelly (2006) has pointed out that the political meaning of emotions emerges through the specific time-spatial context and its embedded power-relations. As the emotional geographies of Ecuador's local political spaces date back to, address, and challenge (post-)colonial injustices, this research is interested in the cultural historicity of these socially constructed emotions. Thus, the book focuses on emotions rather than on the individual biological phenomenon of affect (for a discussion around the differences between affect and emotion see Thrift (2009) and Pile (2009a)). Tolia-Kelly's argument serves as a conceptual starting point for my endeavor to conceptualize the emotional geographies felt, experienced and displayed by differently racialized, gendered, and classed etc. bodies in Ecuador's spaces of politics.

Finally, my endeavor to develop what I call an emotional electoral geography also builds on work in political science which has shown that 'emotions play an explicit and central role in bonding citizens, party, party platform, and elected officials' (Marcus 2002: 37). In his attempt to explain the Bush administration's coalition, political scientist Connolly (2005: 870) argues that 'in politics [...] loosely associated elements [such as affect, spirituality, economic interests etc.] fold, bend, blend, emulsify, and dissolve into each other, forging a qualitative assemblage resistant to classical models of explanation'. Taking Connolly's argument as a starting point, I engage with the emotional and spiritual dimensions of politics in Ecuador. While political scientists have explored the emotional dimensions of campaign advertising (Ansolabehere and Iyengar 1995), campaign speeches (Hampton 2009), campaign events (Van Zoonen 2005), and canvassing (Goldstein and Holleque 2010), their experimental or neuropsychological studies have been mainly concerned with the impact of candidates' emotional performances on electoral results (see e.g. Marcus et al. 2006, Marcus and Mackuen 1993, Marcus, Neumann, and MacKuen 2000). In developing what I call an emotional electoral geography, I aim to broaden this narrow focus towards a fuller engagement with the way emotions constitute spaces of politics in general and spaces of campaigning in general in the first place. Arguing that electoral geographies have to take emotions seriously in order to understand electoral dynamics of populist campaigns in Latin America, this study engages with emotions in two different ways. Chapter 5 develops a performative approach to emotions which focuses attention on the way emotions circulate between bodies, questioning how certain emotions come to stick to certain bodies as outcome of histories of encounter. I argue that such a performative understanding of emotions – as developed by Ahmed (2004a,b) – helps to capture how emotional campaign practices performatively generate the gendered, racialized and classed boundaries of *el pueblo* in different ways in particular times and places. Chapter 6 employs the concept of intersectionality to further understand how (collective) emotional performances within electoral campaigns result from, address and challenge colonial and post-colonial experiences of racism, patriarchalism and classism.

Step 3 on the way towards feminist electoral geographies			
Theoretical lens	Conceptual contribution	Empirical foci	Methodological implications
Emotion and affect (Ahmed 2004a,b, Thrift 2004, Tolia-Kelly 2006)	New way of theorizing political processes that challenges rationalist approaches in electoral geography through understanding emotions, affects and spirituality as central elements that shape political subjectivities, spaces and actions.	Emotions performed, displayed, felt and circulated in spaces of politics	Take into account both linguistic and embodied emotional performances of political subjects' identity performances through interviews, observation and visual methods.
	Analyzing the performativity of emotions serves to capture the genealogies of power saturated spaces of politics.	The emotional histories of encounter between different bodies	Identify emotional patterns in histories of encounter through long-term observation or ethnography, media analysis and archival work.
	Understanding emotional performances as outcome of and mean to challenge intersecting system of oppressions reveals relations between power, politics and identity.	Relations between emotional performances and political subjects' identities	(Visual) ethnography as a method to grasp complex and often subtle relations between identity, emotions and power.

Table 4: Approaching feminist electoral geographies through emotions
Source: Own compilation

(Spaces of) Politics at the 'Crossroad' of Gender, Ethnicity, Class and Locality

The title of this section makes a direct reference to the concept of intersectionality as introduced by Kimberlé Crenshaw (1989). Employing the metaphor of a street crossing, Crenshaw looked at the various ways in which Black women stand at the crossroad of gender *and* race discrimination (Crenshaw 1989, 1993, 1994b, 1995). From a legal perspective, Crenshaw was initially interested to look at the way race and gender interacts to shape the multiple dimensions of Black women's employment experiences. Her metaphor has been critiqued for suggesting that the categories of gender and race exist *before* and *after* the crossroad independently from each other, even though gender and race rather need to be understood as interdependent categories that constitute each other (Becker-Schmidt 2007, Davis 2008, Degele and Winker 2007, Hancock 2007b, Klinger, Knapp, and Sauer 2007,

Walgenbach et al. 2007). While I agree with this critique, I still think that the concept of intersectionality as first articulated by Crenshaw is an important concept for feminist political geographies. Crenshaw basically uses the concept in three different ways in her writings:

1. *Structural* intersectionality: framing of Black women's experience at the intersection of interlocking systems of oppression such as racism, sexism, and homophobia (Crenshaw 1994b, 1995).
2. Intersectional *experiences*: focus on Black women's experiences as shaped by structural intersectionality (Crenshaw 1989).
3. Intersectionality as an approach for *identity politics* focusing on multiple identities (Crenshaw 1995).

Hence, to engage with intersectionality in electoral geography means to look at the intersections of both identities and structures that constitute and shape the spaces of democracy. Acknowledging the difficulties that emerge through the travelling of concepts to other geographic and academic contexts (Knapp 2005, for a discussion why Latin American feminists have not (yet) engaged with the concept see Zapata Galindo 2013), I argue that intersectionality is a productive tool to approach issues of power, inequality and injustice in post-colonial contexts (for the relation between intersectionality studies and postcolonial studies see Kerner 2010).

Further, I consider the concept of intersectionality an appropriate tool to engage with the distributional dimension of politics, the third approach in Staeheli and Kofman's (2004: 2) systematization. Focusing on the distribution of power, resources and privileges, this approach is mainly interested in the spatial patterns of inequality as a way of demonstrating the exercise of power. When looking at the spatial patterns of inequality, it has to be asked who is included or excluded in processes of decision-making, who has access to the spaces where decisions are taken, who benefits from the way resources are distributed and on what grounds. Intersectionality helps to analyze these questions in a very differentiated way through focusing on (1) the way different structures of oppression intersect in shaping the distribution of power, privileges, and resources in particular spaces and historical moments, (2) through engaging with the way the distribution of power and resources affects the experiences of people due to their intersectional positioning, and (3) how a hegemonic order that defines the distribution of power and resources is contested by different social groups and their intersectional identity politics. Given the strong fit between questions that feminist political geographers raise and answers that the concept of intersectionality promises to provide, it comes as no surprise that feminist debates on the intersection of multiple axes of power and difference have profoundly shaped the emergence of feminist political geography (Nelson 2006: 369). The concept offers the possibility to pay

> 'explicit attention to women, gender and sexuality, and the ways in which other axes of identity are entwined with these in the relationships of power, oppression and domination that organize and construct the social world' (Brown and Staeheli 2003: 247).

I would like to push the argument further by arguing that the concept of intersectionality is not only central for a feminist political geography, but can also make fruitful contributions to feminist electoral geographies.

In analogy to Crenshaw's three dimensions of intersectionality, I identify three ways in which the concept of intersectionality sheds light on questions of the distribution of power, resources, and privileges in electoral politics: First, intersectionality forces us to identify the multiple barriers to power like sexism, racism, and other forms of exclusion and to ask how these systems interrelate in the constitution of spaces of politics. These structural intersectionalities can be identified both by looking at the histories and genealogies of electoral politics in a certain context (see previous section) and by analyzing electoral results from an intersectional perspective. Engaging with the histories and genealogies of the spaces of democracy in a certain context makes the systems of exclusion visible on which democracy has been established in this particular place (see chapter 3). By identifying these historical exclusions, we can trace whether these intersectional structures persist until today or have been subject to change. The analysis of the development of electoral results provides information about the exclusion and inclusion of particular gendered, racialized, and classed subjects at different historical moments (see chapter 7). When looking at electoral results from a historical and intersectional perspective, it is especially important to be attentive to those subjects that do not appear in the electoral statistics and to ask how their exclusion is a result of intersecting systems of oppression. Further, the intersectional analysis of campaign budgets, performances on the electoral stage (who comes to speak how long in what place?), and orders on the party lists (who is positioned first on the list?) further sheds light on the way electoral politics is gendered, racialized, sexualized, and classed. To identify the structural intersectionality that prevents certain gendered, racialized, sexualized, and classed subjects from participating in electoral politics is a first step to create a more inclusionary polis. An intersectional analysis is a crucial requisite for thinking about and designing policies that promote the electoral participation of excluded social groups.

Second, the lived experiences of both privileged and marginalized political subjects within a particular time-spatial context further reveal the consequences of the structural intersectionalities that constitute the spaces of democracy. Along the example of a deaf woman, Valentine (2007) has shown how conceptualizing intersectionality as a 'lived experience' serves to understand power relations as operate in and through the spaces within which we live. These place-specific power relations shape the processes of identification and dis-identification of individuals through the defining hegemonic cultures that dominate in particular spaces. For Valentine (2007: 15), such an analysis means asking questions about

> 'what identities are being "done", and when and by whom, evaluating how particular identities are weighted or given importance by individuals at *particular moments* and in *specific contexts*, and looking at when some categories such as gender might unsettle, undo, or cancel out other categories such as sexuality'.

Valentine's approach not only highlights the importance to consider the particular time-spatial context in which processes of identification are embedded, but further calls for the necessity to think space and identities as 'co-implicated' (Valentine 2007: 19) in the conceptualization of intersectionality (for examples how this approach has been employed in geography see Büchler 2009, Herzig 2006, Herzig and Richter 2004, Schurr 2011, Schurr and Segebart 2012, Schurr and Wintzer 2012). To think intersectionality in such a way as a lived experience facilitates an understanding of the 'intimate connections between the production of space and the systematic production of power' (Valentine 2007: 19). Employing Valentine's approach within the field of electoral geography offers the possibility to bring together the (intersectional) enactment of identity within the spaces of democracy and the social structures of patriarchy, racism and heteronormativity that enable and discipline the lived experiences of the political subjects within these spaces. In this way, identity is conceptualized rather as contingent on the power saturated spaces in and through which the political subjects' experiences are lived than as a multiple and fluid product of intentional, conscious agents. Hence, by researching identity performances on the political stages through the lens of intersectionality, we can learn more about the processes of inclusion and exclusion that constitute the spaces of democracy in different moments and places in different ways. To understand how people experience the exclusionary structures of electoral politics in an empirically grounded way is an important step in tackling these structures through the design of inclusionary policies.

Third, following Crenshaw's understanding of intersectionality as a ground for identity politics, I employ the concept to evaluate policies and political action. I address the question in what ways policies aim to redress structural intersectionalities by drawing on discussions in political science around the intersectionality of equality policies (Borchorst and Teigen 2009, Hancock 2007a, 2007b, Simien 2007, Townsend-Bell 2011). Political scientists have engaged with the question of how affirmative policies such as quotas or reserved seats have succeeded (or not) in overcoming structural intersectionalities with regard to the access of marginalized groups to the spaces of electoral politics. They have shown that affirmative policies such as the quota that tackle the problem in a unitary way by addressing discrimination based on one single identity category – mainly gender or ethnicity – often fail to promote the participation of, for example, minority women. While quotas have been successful in raising the number of female candidates in electoral politics in general, they benefit majority women (e.g. white women) more than minority women (e.g. indigenous or colored women) (Hughes 2011). Building on these findings, feminist electoral geographers are challenged to think about affirmative policies that take into account the intersectionality of systems of exclusion. Asking what different systems of exclusion intersect in a particular space of democracy (such as a municipality in the Andes highland) due to its particular history, more differentiated, situated policies can be developed that are 'sensitive to the constitutive role of spatiotemporal context' (Peck 2011: 773). This third approach, however, is not only interested in thinking about affirmative policies that promote the participation of marginalized groups in electoral politics, but also in policies and political actions that aim to address social inequalities more broadly. In my work, I analyze the policies designed and launched by new political subjects from an intersectional perspective in order to evaluate whether they design their policies on the ground of a unitary or intersectional identity politics (see chapter 7). Such an evaluation offers the possibility to analyze who benefits from the policies launched by new political subjects or whether these policies produce new exclusions and marginalizations.

To sum up, feminist geographers have been concerned with questions of intersectionality (often without using the term intersectionality itself) from the beginning of feminist geographies (for early accounts see Kobayashi and Peake 1994, Peake 1993, Pratt 1999, Ruddick 1996). A feminist electoral geography can build on this body of work, when looking at the intersectionality of structures, identities and policies that shape the geographies of elections and politics. In so doing, the power relations that saturate the spaces of democracy can be better understood in terms of how systems of exclusion intersect in specific ways within the spaces of politics due to their particular history. Intersectionality further helps to understand how the lived experiences of political subjects are shaped by these exclusionary structures, and how these structures are contested by policies seeking to address social inequalities.

Step 4 on the way towards feminist electoral geographies			
Theoretical lens	Conceptual contribution	Empirical foci	Methodological implications
Intersectionality (Davis 2008, McCall 2005, Valentine 2007, Winker and Degele 2011)	Identify power relations that saturate spaces of politics and their transformation across time and space.	Intersectionality of structures that define access to electoral politics	An intersectional analysis of electoral results provides information about structural in-/exclusions in politics.
	Political subjects as consisting of multiple identities that shape their access to the spaces of politics.	Intersectionality of identity constructions of political subjects	Intersectional analysis of interviews, fieldnotes and visual data.
	Focus on the intersectional impact of policies (substantive representation) rather than only the descriptive representation of women	Intersectional policies	An intersectional analysis of political agendas asks who benefits from certain policies.

Table 5: Approaching feminist electoral geographies through intersectionality
Source: own compilation

Rethinking Electoral Geography through Feminist Theories

In the previous sections, I have presented three different approaches to rethinking electoral geography through a feminist lens. At this point, I would like to briefly summarize the key points of my theoretical perspective. Aiming to challenge the lacking theoretical engagement of more conventional electoral geographies (Barnett and Low 2004, Johnston and Pattie 2004), my approach has aspired to rethink the central concepts in electoral geography by drawing on feminist (political) theories. In short, the theoretical framework suggested throughout this section combines Mouffe's understanding of politics as antagonism, Butler's notion of performativity, and Crenshaw's concept of intersectionality in order to reconceptualize 'politics'/'the political', power, space, identity, and difference as key concepts for a feminist electoral geography. The three perspectives complement each other by approaching each of these concepts from different theoretical angles. Bringing them into dialogue offers the possibility to ground electoral geography in feminist theory. To conclude, I briefly sketch how my theoretical approach advances a theoretical foundation of electoral geography:

Rethinking the binary between politics and the political: Employing Mouffe's approach that differentiates between 'politics' and 'the political', I have shown that it is crucial to think electoral politics not in a binary way as opposite to non-institutionalized political action, but to think of both spheres as co-constitutive. Understanding 'politics' as a set of practices and institutions that organize and order the antagonistic relations of 'the political' turns the institutions of (state) politics into an object of contestation. Thereby, the established hegemonic order is constantly challenged through counter-hegemonic imaginaries or horizons of what this order should look like. Thinking elections and electoral geography more broadly through this theoretical lens brings the relationship between electoral practices – such as assigning candidates, campaigning and voting – and antagonistic struggles articulated outside the spaces of institutionalized politics such as social movement protest to the center of electoral analysis. Hence, electoral results need to be analyzed against the backdrop of antagonisms identified in specific electoral territories. In so doing, the political action of those political subjects whose activism is located outside the spaces of state politics is taken into account in electoral research that has long been neglected by 'conventional' electoral geography.

Re-locating politics: The endeavor to reconceptualize the relationship between electoral politics and the political action of social movements goes along with re-locating politics in both public *and* private spaces. Taking into account spaces that are not so obviously constructed or maintained by the state is crucial for a feminist electoral geography that seeks to understand the relation between electoral politics and the political and aims to identify new spaces of democracy that are constituted outside the sphere of the state. A performative understanding of space as articulated by Gregson and Rose (2000) demands to center on the (power-saturated) practices, performances and interactions that constitute political spaces/spaces of electoral politics. Focusing on the practices of political subjects such as politicians, electoral candidates or constituents and following them in their everyday lives enables us to identify spaces of/for democracy beyond the spaces officially assigned to state politics such as the (building of the) municipality, provincial council, national assembly. Further, by studying the citational practices that constitute the performative spaces of politics, it is possible to trace the genealogies of the spaces of democracy. Paying attention to the small slippages that occur in the moment of reiteration, changes become visible in the way these spaces are brought into being through slightly different practices, interactions or material constructions. A feminist electoral geography benefits from such a performative perspective as it brings agents of change to the center of attention that are often overlooked, such as changing legislations, electoral laws, divergent campaigning practices, new technologies etc.

Re-conceptualizing identity: Identity has long been on the agenda of electoral geography, be it regional identity of electoral territorialities, gender, ethnicity, or religion. Identity, however, has mainly been conceptualized as static, socio-demographic data, as variables of analysis that explain electoral results rather than as category that needs further inquiry itself. Feminist (poststructuralist) theorists have challenged this static notion of identity by highlighting the contingent, fluid, and multiple characters of identit*ies* (the plural is programmatic here). While the cultural turn has successfully integrated poststructuralist thinking in human geography in general, electoral geography has been somehow unaffected by the poststructuralist turn. I argue that Butler's theory of performativity and the concept of intersectionality are important tools to integrate a poststructuralist notion of identity in electoral geography. While Butler's thinking sheds light on the way political practices have historically shaped masculinity and femininity as a central characteristic of (a-)political subjects, intersectional approaches capture the relation between the intersecting structures of exclusion that define access to political power and the intersectional experiences of people that are shaped through these structures.

Re-placing power: Power is at the center of the three discussed theoretical approaches. For Mouffe, questions of power (or hegemony, as she prefers to call it) arise through antagonistic struggles over the established (hegemonic) order. As this order (created and maintained through 'politics' and its institutions and practices) can be considered the expression of a particular structure of power relations, it is always based on some form of exclusion. Hence, questions of who is included and who is excluded are central to an antagonistic understanding of 'politics' and 'the political'. Butler, in a similar vein, is interested in de-naturalizing the sedimented established norms by revealing through her performative approach the genealogies of gender or political norms. Considering the expansion of these norms to 'previously disenfranchised communities' as a central 'task of a radical democratic theory and practice' (Butler 2004b: 225), Butler urges us to question and transform the power relations that saturate the spaces of democracy. The concept of intersectionality as developed by Black feminists (Collins 1999, Combahee River Collective 1977, Crenshaw 1989, Davis 1981) offers the possibility to do so: First, by analyzing how power structures intersect within a particular place due to the specific history of this place. Second, by engaging with the way these intersectional structures shape the political opportunities of people in a particular place at a particular moment. And third, by addressing social inequalities and promoting the access of marginalized social groups to the spaces of democracy through intersectional policies. Taking questions of power, exclusion, and inclusion seriously, is crucial for feminist electoral geographies seeking to contribute to a more just and inclusive democracy. The three suggested approaches all offer the possibility to re-position power at the center of empirical analysis in electoral geography.

Having laid out a theoretical framework to re-think electoral geography, the next section asks how these theoretical perspectives can be redeemed methodologically.

STUDYING POLITICAL PRACTICES IN LOCAL SPACES OF POLITICS

> 'Geographers are well placed to discuss the theoretical foundations and development of formal and informal electoral institutions, [...] the gendered, racialized and sexualized geographies of elections, and post-structural perspectives that see elections as embodied political performances that produce particular societal discourses and effects. The shifted focus onto critical and post-structural election studies likewise suggests that greater use of qualitative methods may help uncover new understandings of electoral processes as grounded, contingent, and embodied processes deeply intertwined with the broader societal milieu' (Nicley 2011: 76).

> Feminist political geographers should follow the 'call for close ethnographic attention to the local construction of women's political participation' (Secor 2004: 271).

The two introductory quotes call for a shift with regard to the methodologies employed in electoral geography, advocating the potential of qualitative and ethnographic methodologies for poststructuralist/feminist electoral geographies. Following these calls, I present in this section my methodological framework, which combines quantitative and qualitative methods. I argue that a multi-method approach is suitable to bring together electoral results with broader social structures and political subjects' lived experiences. Outlining the methodological approach employed in my research on women in Ecuadorian local politics, I aim to contribute to the development of new feminist methodological approaches to electoral geographies.

Considering collaboration a keystone of feminist methodologies (McDowell 1992, Monk, Manning, and Denman 2003, Moss 2003, Nagar 2013, Nast 1994, Sharp 2005), my research interest and the resulting research questions were inspired through conversations with women politicians in the local rural governments *(Juntas Parroquiales Rurales,* JPR*)* and with representatives of the National Organization of these local rural governments *(Consejo Nacional de Juntas Parroquiales Rurales del Ecuador,* CONAJUPARE*)* (see also chapter 7). It was when working as an intern for the German Development Cooperation (GIZ) and writing my Master thesis about one of their training program for members of the newly founded rural local governments (Schurr 2009a) that I first engaged with the question of how these new institutions transform or reproduce power relations that saturate local communities. During this first project, women politicians discussed with me the paradoxes they experience in their everyday life, trapped between a progressive legislation that obliges parties to include women on their lists and persisting discrimination against women in local politics. My PhD project resulted from these conversations and the women's call for more (detailed) information about the situation of women in local politics in order to be able to tackle the obstacles these women face. At the same time, the CONAJUPARE offered me the possibility to collaboratively conduct a research project on the situation of women, ethnic minorities and young people in the local rural governments (JPR) (Mosquera Andrade, Schurr, and CONAJUPARE 2009). In a collaborative process, we developed a research design that combined a quantitative questionnaire sent out to all 798 local governments throughout the country and thirty in-depth interviews with selected local politicians of the JPR on the basis of a representative sampling. Given that the information provision of electoral data is rather poor in Ecuador, the study provided unique data as it specified the gender, age, ethnic self-identification and political experience of local politicians (for an analysis of the collected data see chapter 7). On basis of the complete sample three case study areas were selected. The selection was based mainly on the two following criteria: the presence of female, indigenous and Afro-Ecuadorian local politicians in leading executive positions (basically prefect and mayor) and regional differences that are decisive for the ethnic composition of Ecuador's provinces. The following map shows the participation of women in local electoral politics in the three case study areas in the provinces of Esmeraldas, Chimborazo, and Orellana.

Carrying out research in three case study areas – each of them having its particular (post-)colonial history, different socio-demographic characteristic, and diverging level of female participation in electoral politics – offered the possibility to engage with the way gender and ethnicity intersect in different places in different ways. To get a sense of the messy antagonistic and intersectional power relations that structure and shape each of the political territories, I spent a total of eighteen months living and researching in the provinces of Esmeraldas, Chimborazo, and Orellana between 2006 and 2010.

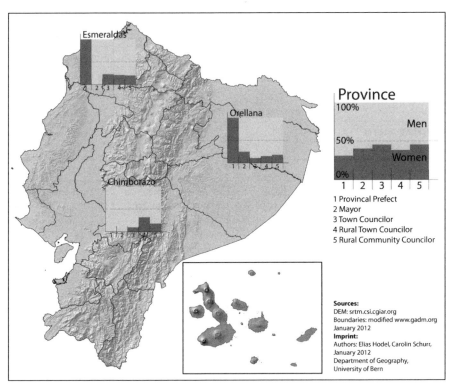

Map 1: Participation of women in electoral office in the three case study areas
Source: Elaborated by the author on basis of data provided by CONAMU 2009

The main socio-demographic characteristics of each province, the elected women and ethnic minorities as well as the interviews conducted in each of the provinces are described in the following table:

	Province Esmeraldas	**Province Orellana**	**Province Chimborazo**
Region	Coast	Amazon lowland	Highland
Ethnic composition[1] (self-identification)	3 % indigenous 45% mestiza 44% Afro-Ecuadorian	31 % indigenous 57 % mestiza 5 % Afro-Ecuadorian	38 % indigenous 58 % mestiza 1% Afro-Ecuadorian
Level of poverty[1] (under 2 USD/day)	78% of population	85 % of population	67 % of population
Illiteracy[1]	10 %	6 %	13%
(post-)colonial history	afro-descendant community as free slaves; absent post-colonial state	extraction of precious wood, rubber and oil; absent state in both colonial and post-colonial times	colonial *hacienda* system; indigenous tribute until 1857
Presence of social movements	center of Afro-Ecuadorian movement	anti-petroleum, environmental and human rights movements	center of indigenous uprising
Women in executive positions in local politics	1 prefect, 1 vice-mayor	1 prefect, 1 mayor	1 vice-prefect
Ethnic minority people in executive positions in local politics	1 Afro-Ecuadorian male mayor 1 Afro-Ecuadorian female vice-mayor	1 indigenous male mayor	1 indigenous male prefect 3 indigenous male mayors
Interviews conducted (all interviewees are female unless stated otherwise) Age: 20–68 Education: ranging from 2 years of primary to university degree Political parties: Pachakutik (17), Amaute (3), MPD (13), PRIAN (3), PSP (5), Alianza Pais (6), PSC (3).	1 Afro-Ecuadorian *Asambleista* 1 Afro-Ecuadorian prefect 2 vice-mayor (afroec. & mestiza) 5 town councilor (1 afroec., 4 mestiza) 1 male Afro-Ecuadorian president of local rural government 5 rural government councilors (2 afroec., 3 mestiza) 5 women from women's organizations	1 mestiza *Asambleista* 1 male indigenous *Asambleista* 1 mestiza prefect 1 mestiza mayor 5 town councilor (1 Afro, 4 mestiza) 1 indigenous president of local rural government 5 rural government councilors (3 indig., 2 mestiza) 4 women from women's organizations	4 *Asambleistas* (2 indigenous, 2 mestiza) 1 male indigenous prefect 1 indigenous vice-prefect 6 town councilor (2 indigenous, 4 mestiza) 4 rural government councilors (2 indig, 2 mestiza) 6 women from women's organizations

Table 6: Information on case study localities and interviews conducted
Source: [1]SIISE and own compilation

Accompanying local female politicians in their everyday routines, attending council sessions, office hours, and internal meetings, I tried to learn as much as possible of what it is like to 'do' politics in each of the three case study areas. In each province, interviews were conducted with female politicians elected at the parish, municipal and provincial level. In total, semi-structured, partly biographical interviews were conducted with 43 local female politicians and 15 women from local women's organizations. When the elections of 2009 were approaching, I accompanied female candidates of the party *Pachakutik* in Orellana and Chimborazo and female candidates of the socialist party MPD *(Movimiento Popular Democráctico)* in Esmeraldas, spending a week in each of the case study localities during the three weeks of electoral campaign. As a result of the new Constitution passed in 2008, all electoral offices on all political levels were newly elected in 2009 – turning this election into a 'super election'. Further, it was the first election in which the gender quota law reached parity, requiring equal numbers of women and men placed in alternation on each party list.

Especially during the campaign events, but also during other political activities, I conducted a visual ethnography (see chapter 4). This (visual) ethnographic approach (Pink 2001, 2007, 2009) facilitated a stronger analytical focus on the embodied practices, emotions, and interactions that constitute the performative spaces of politics. The photographs and videos recorded during fieldwork were later used to (re-)construct a close description of events, focusing on what the candidates' and constituents' bodies were doing, their facial expressions, their interactions, and their gestures. In so doing, I aimed to capture the grounded, emotional, and mundane practices that constitute and shape the spaces of politics, which are often overlooked in more conventional qualitative research such as interviewing (a similar call for ethnographic approaches in political geography has been made by Kuus 2012, Megoran 2006, Woon 2013). Conducting (visual) ethnography of elections, I was interested in the performances on the political stages, the interactions between politicians and the audience, and the emotions circulating in these spaces of campaigning. Hence, in contrast to more conventional electoral geography that focuses on the analysis of electoral results, I was concerned with the practices that constitute the spaces of elections in the first place. The videos recorded during the campaign now serve to transmit an impression of the observed events; some of these recording are presented to the reader in linked videos (see chapter 4 and 5).

After the elections, once the results were published, I analyzed the electoral results of the parish, municipal, and provincial elections both in the three case study areas and the rest of the country. In my analysis, as far as the data permitted, I was especially attentive to the way gender, ethnicity, class, and locality (rural/urban) intersected in the electoral outcomes (for more information see chapter 7).

Table 7 shows how each of the applied methodological approaches contributes to the development of a new methodological framework for a feminist electoral geography.

Step 5 on the way towards a feminist electoral geography			
Methodology	Empirical data	Empirical foci	Contribution
Intersectional quantitative analysis (McCall 2001)	Intersectional analysis of local and national elections	Electoral results: in-/exclusions in the electoral participation of social groups	Focus on questions of power, ex-/inclusion and their spatial patterns when studying electoral results
Interviews (Degele and Winker 2007)	Interviews with women politicians and representatives of women organizations	Intersectional lived experiences Political biographies Linguistic identity performances	Grasp everyday experiences/struggles of local politics through the lens of political subjects
Ethnography (Megoran 2006)	Ethnographic notes from accompanying politicians in their activities	Identify power relations, routines, gossip, agendas, discrepancies between what is said and what is done	Grasp everyday routines and citational practices that (re-)produce the hegemonic order of political institutions
Visual ethnography (Garrett 2011, Pink 2001, Rose 2011)	Videos of electoral campaign and political events, collection of visual representations	Research emotions, embodiment, interactions, performances that constitute the spatiality of politics	Focus on the construction of political spaces and identities as performative, emotional and intersectional

Table 7: Suggested methodological research agenda for a feminist electoral geography
Source: Own compilation

Returning to my theoretical framework, I conclude this section by summarizing how the applied methodology redeems the theoretical premises presented in the previous section 'Theorizing Spaces of Democracy'.

(Re-)searching power structures: Power is placed at the center of a feminist engagement with politics, no matter whether approached through the theoretical lens of antagonism, performativity or intersectionality. What makes all three approaches so productive for a feminist electoral geography is their focus both on broader societal power structures such as racism, sexism, heteronormativism and the micro-power in the sense of Foucault that saturate places in multiple ways (Foucault 2005, Massey 1993, 1999): Antagonisms can be identified both in terms of the dominating political power structure between national government and opposition or colonizers and colonized for example *and* in terms of a temporary, place-specific conflict within a local government. The concept of performativity connects the citational practices that produce discourses with the performances of subjects that (re-)produce and contest these discourses. Finally, intersectionality links the interlocking systems of oppression with the lived experiences in particular places. To grasp the relation between structural power relations and subjective experiences, I have employed a range of methods, each of them identifying power

relations on different analytical levels. Analyzing electoral results and socio-economic data (rates of poverty, illiteracy, income, etc.) from an intersectional perspective, I focused on social inequalities on a structural level. In contrast, interviews with women in formal and informal politics in each of the three case study areas and extended ethnographic observation engaged with the way people (intersectionally) experience and (performatively) (re-)produce and /or challenge power relations (including structural power relations) in their everyday life in this particular place. Bringing together the identified structural inequalities and the place-specific (micro-)power relations, I hope to understand on what grounds spaces of politics are constructed as exclusionary and to what extend some spaces of politics have been transformed into more inclusive spaces of democracy.

(Re-)presenting political identities: All three theoretical approaches are further interested in understanding how identities are constructed: Antagonism focuses on the construction of collective political identities (us/them), performativity engages with the constitution of gender (and racial) identities, and intersectionality conceptualizes the relationship between different identity categories. Qualitative (case) studies, interviews, and ethnographic observation have been at the center of the methodological discussions in feminist geography about how to engage with identities in a poststructuralist way (Andolina, Radcliffe, and Laurie 2005, Bowleg 2008, Gibson-Graham 1994, McDowell 2008, Nelson 2006, Peake and Trotz 1999, Valentine 2007). In my research, I have employed interviews to capture the (linguistic) performances through which the women interviewed construct their identities. Observing the women while accompanying them in their everyday lives, I could witness how certain identities become more important in certain contexts and become irrelevant in others (for results see chapter 7). Finally, by conducting a visual ethnography (see chapter 4), I was able to capture the embodied and linguistic performances through which the female politicians constructed their identities in different ways on different political stages. Analyzing the emotional, embodied and linguistic performances of the observed political subjects has served to evaluate to what extent the new political subjects like women or ethnic minorities actually perform political subjectivity differently, 'do' politics differently and hence transform or decolonize the post-colonial spaces of politics.

(Re-)constructing spaces of politics: While Mouffe, Butler and Crenshaw address space only implicitly, human geographers have incorporated their work in discussions about spatial thinking (Antagonism: Barnett 2004, Performativity: Cloke, May, and Johnsen 2008, Gregson and Rose 2000, Massey 1995, Intersectionality: McDowell 2008, Nightingale 2011, Valentine 2007, 2008). In different chapters, I discuss how an antagonistic (chapter 3), performative (chapter 3,4,5) and intersectional (chapter 6,7,8) conceptualization of space contributes to identifying, analyzing and understanding processes of political transformation. But, how can we methodologically capture the spatial traces of these processes? In chapter 4, I suggest visual ethnography as an appropriate methodology to capture the minute and mundane interactions, linguistic and embodied performances that bring into being

the performative spatialities of politics inside and outside the material buildings of state institutions. Chapter 6 shows how a (visual) ethnography of the emotions displayed and articulated in the spaces of campaigning sheds light on the way these spaces are gendered, racialized, and classed. While these two chapters focus on the relational, performative dimension of space, chapter 7 engages more directly with the material traces of women politicians' activism. Long-term observation and interviews enabled to reconstruct how the material construction of the spaces of politics has changed (or not) since new political subjects inhabit these spaces.

PERFORMATIVITIES

3.
THE PERFORMATIVITY OF POST-COLONIAL POLITICS

'Anyone trying to mesh theory with empirical description soon learns that the movement among abstract concepts and empirical description is like performing ballet on a bed of quicksand' (Pudup 1988, 384).

TAKING BUTLER AND MOUFFE ELSEWHERE

'Surprisingly enough, relatively little critical reflection has been devoted to what constitutes the proper political domain, to what and where is "the political"' (Swyngedouw 2008: 3).

'This political space I am occupying is a space from which women and especially *Shuar* women have always been excluded' (interview with an indigenous national deputy of the Ecuadorian *Asamblea Nacional*).

Reading the interview quote from the *Shuar* deputy against Swyngedouw's reflection, we have to ask not only where the political is located but also who is included or excluded in the spaces of politics. In this chapter, I engage with the question how spaces of politics can be conceptualized so that the power relations and resulting hegemonies and acts of marginalization that constitute these very spaces become visible. Departing from Gregson and Rose's performative approach to space, I suggest that their notion of performative space is inherently political as it situates power relations at the center of their understanding of space. At the same time, I want to push Gregson and Rose's argument further by elaborating conceptually on the antagonistic power relations that constitute the spaces of politics. By bringing Mouffe's notion of politics as antagonism/agonism into dialogue with the concept of performative space, I intend to shed light on how spaces of politics are brought into being, materialize hegemonic political orders and are subject to contestation.

The fictitious dialogue between Butler – via Gregson and Rose – and Mouffe to be developed in this chapter aims to contribute to ongoing discussions in political geography about the relations between politics/political, space and social change by addressing the following questions: How are spaces of politics brought into being through regulatory, citational practices and performances? How are hegemonic spaces of politics reproduced, contested, resisted and changed by counter-hegemonic political subjectivities? What kinds of spaces of politics result from politics of antagonism and how could an agonistic space of politics look like?

I think through these questions by drawing on my research that deals with the increasing presence of new political subjects in Ecuador's electoral politics. My ethnography of local politics looks at the way local spaces of politics are brought into being by the citational practices of new political subjects and questions whether the hegemonic (post-)colonial mode of doing politics is challenged by these practices.

The chapter starts with a brief summary of Gregson and Rose's argument before the central scholarship in political geography that engages with questions of performativity is revisited. Then, highlighting the shortcomings of a performative approach to spaces of politics, I suggest bringing into dialogue the notion of performative space with Mouffe's understanding of politics as antagonism/agonism. I think through this theoretical dialogue between Mouffe and Butler – via Gregson and Rose – along my empirical research about transformations in local spaces of politics in Ecuador in three steps: First, drawing on the (post-)colonial history of Andean politics, I ask how a spatialized political order is established and maintained. Second, focusing on social movement struggles in Ecuador's recent history, I highlight how Butler's notion of subversion and Mouffe's concept of disarticulation are suitable to frame counter-hegemonic struggles that contest the hegemonic political order. Third, taking up Mouffe's idea of agonistic pluralism, I discuss whether the Ecuadorian political imaginaries of *interculturalidad* and *plurinacionalidad* could actually expand the categories of the political towards an agonistic democracy. Finally, I turn back to Gregson and Rose's performative thinking of space, asking how their conceptualization of space can be turned into a vital tool for political geography.

THINKING SPACE THROUGH PERFORMATIVITY AND ANTAGONISM

'Despite the growing interest in performativity among critical geographers, this work has not made many inroads within mainstream political geography' (Rose-Redwood 2008: 879).

In 2000, Nicky Gregson and Gillian Rose (2000) wrote in their seminal paper 'Taking Butler elsewhere: performativities, spatialities and subjectivities' that so far, geographers had rather drawn on Goffman's account of performance than on Butler's understanding of performance and performativity to highlight that social life resembles some sort of performance. In the meantime, Butler's concept of performativity has turned into an important source of inspiration for geographers working on issues as diverse as bodily practices (Longhurst 2000), questions of (gendered, sexualised and racialised) identity (Hubbard 2000, Mahtani 2002, Thomas 2008), homelessness in the city (Cloke, May, and Johnsen 2008), identity performances in workplaces (McDowell 2008, 2009) and the performativity of research itself (Gregson and Rose 2000, Pratt 2000a). By now, 'Butler's model of performativity has provided food for thought for many geographers' (Mahtani 2002: 427) in human geography. At the same time, however, Butler's work has been critiqued for making 'little room for space' (Thrift and Dewsbury 2000: 414). A gap Gregson and Rose (2000) started to fill when developing their performative approach to space which is based on Butler's notion of performativity. By framing space in terms of power when arguing that 'space too needs to be thought of as a performative articulation of power' (Gregson and Rose 2000: 434), they conceptualize space in an inherently political way.

It is therefore surprising that their notion of performative space has received little attention in political geography. In general, the engagement with performativity is still rather scarce within political geography in contrast to a more profound involvement with the concept in feminist and gender geographies (Bell et al. 1994, Bell and Valentine 1995, Longhurst 2000, Nelson 1999b, Pratt 2004, Secor 2003), social and cultural geography (Cloke, May, and Johnsen 2008, Dewsbury 2000, Malbon 1999, Nash 2000), and economic geography (McDowell 2008, 2009). Within political geography, performativity has been employed to approach the political genealogy of scales (Kaiser and Nikiforova 2008), political identities (Kuus 2007), questions of citizenship (Baird 2006, Mahtani 2002), borders (Strüver 2005b) and political toponomies (Rose-Redwood and Alderman 2011, Rose-Redwood 2008).While all these accounts engage in the relation between performativity and the constitution of the political, the very spatialities of politics and the way these spatialities are (re-)produced or contested have been left unaddressed.

In my view, Gregson and Rose's (2000) account of performative space appears to be a productive way to conceptualize spaces of politics as it focuses on the reiterative citation of regulatory practices that constitute and constantly reproduces spaces of politics as hegemonic regimes. What makes both Butler's as well as Gregson and Rose's work especially interesting for a political geography interested in the 'entanglements of power' (Sharp et al. 2000) and related geographies of domination/resistance is their focus on subversion. Butler (1999) highlights that within the reiterative citation of discourse, slippages are possible as there is no guarantee that a repetition is successful; its disciplines may fail. Hence, thinking identities and spatialities as performative 'provides opportunities for radically redoing gender' (Cream 1995: 39), and redoing space respectively. The remark made by Chambers and Carver (2008: 148) that it is not identity (or space) itself which must be subverted, but the regulatory practices that discipline and enable certain identities (and spaces), is crucial when thinking about political change as it sheds light on the very practices that bring (counter-)hegemonic identities and spaces into being within contexts of transformation. This observation is also essential in regard to the discussion about agency in Butler's work (Nelson 1999b), because it highlights that Butler (1993: xxiii) locates agency in a Foucauldian manner in the reiterative and articulatory practice immanent to power, and not as a relation of external opposition to power (Foucault 2005). Based on this understanding of agency, Gregson and Rose (2000: 441) have highlighted that it is vital for a critical geography concerned with the constructedness and provisionality of identities, spaces and power relations to 'conceptualize the performers as in some sense produced by power'. While Gregson and Rose have pushed Butler's argument further by arguing that spaces too need to be thought of as performative of power relations, their work gives little hint how to conceptualize these (collective) power relations that are constitutive for any notion of performative political space.

To address this conceptual gap in Gregson and Rose's thinking, I suggest to bring Mouffe's (1993, 2005a) understanding of the political as antagonism/agonism into play with their notion of performative space. By doing so, I seek to shed light on the antagonistic political struggles between hegemonic and counter-hegemonic political subjects that bring the very spatiality of the political into being. For Mouffe (2005b: 18), every political order is the 'temporary and precarious articulation of contingent practices' that have turned hegemonic within a certain temporal and spatial context. The hegemonic articulatory practices that establish and define a certain specialized political order result from an ongoing antagonistic struggle between different collective identities. This hegemonic order is at the same time always 'susceptible of being challenged by counter-hegemonic practices that attempt to disarticulate the existing order' (ibid.). Hence, by understanding the political as antagonism between differently powerful political subjects, Mouffe focuses explicitly on the processes of hegemonialization that are presented in the work of Gregson and Rose rather as a black box.

While Mouffe's work complements the concept of performativity and performative space by focusing on the antagonistic struggle of collective identities, at the same time, performativity helps to address a shortcoming in Mouffe's elaboration of politics as antagonism. Mouffe's politics of antagonism are based on theories of practices; however, she does not give empirical evidence how antagonism is brought into being through everyday political practices. The concept of performativity brings into focus how the political spaces of antagonism are constituted through everyday political practices. I suggest that Butler's work, centering on performativity and Mouffe's work concerned with antagonism have much to say to each other and to a political geography of change.

In fact, there is a major confluence between the works of Butler and Mouffe/Laclau in general. Butler (1993: 146) has drawn on the concept of antagonism for example in 'Bodies that Matter' to talk about the impossibility of a fixed subject position of 'women' and has published a co-edited book with Ernesto Laclau about questions of universality and hegemony (Butler, Laclau, and Zizek 2000), Mouffe in contrast has not directly engaged so far with Butler's work. While (political) geographers have worked both with Mouffe/Laclau (Alsono 1992, Massey 1995, Natter 1995, Pugh 2005, 2007) and Butler (see references discussed above), only few have discussed the parallels between Mouffe/Laclau and Butler mainly in regard to their understanding of discourse (analysis) (Mattissek 2008, Müller 2008, 2009).

My thinking about the performative dimension of political spaces adopts Mouffe's (1995: 262–263, 2005b: 8–9) useful distinction between the political and politics by differentiating between political spaces and spaces of politics. Following Mouffe, I understand political spaces as the spatial materialization of antagonism inherent in all human society. Hence, a political space is any space that is brought into being through the antagonistic relation between hegemonic and counter-hegemonic subjectivities. Political spaces result from a wide set of (everyday) political activities ranging from classic forms of counter-hegemonic political mobilization like public protest to hidden forms of resistance in private spaces (e.g. Scott 1990). In contrast, spaces of politics are the outcome of a 'set of practices and institutions through which an order is created' (Mouffe 1999: 9). In democratic societies, these spaces of politics include the juridical, executive and legislative institutions that constitute the pillars of any democracy. In this chapter, I would like to concentrate on the latter one[8], showing however, that frequently the apparently clear-cut boundary between political spaces and spaces of politics is blurred. By doing so, I aim to overcome the stark opposition found in (feminist) political geography between representative forms of democratic politics, presumed to be the source of dissatisfaction because of its state-centrism, and idealized models of alternative politics (Barnett and Low 2004: 7).

HEGEMONY, POWER AND SPACES OF POLITICS

'Power often comes from claims concerning what is original, primordial, natural, inevitable, factual, genuine, real, scientific and the like, typically in a "founding" or "grounding" narrative of certainty' (Chambers and Carver 2008: 23).

'The spaces of the political have been tightly circumscribed by masculinist ideals. Mouffe's key argument, that the political is an unavoidably adversarial realm, disrupts this superficial neutrality, relying instead on the "affirmation of difference" and on the relational, not rational, nature of identity' (Thien 2007: 134).

I would like to set off with Butler's (2000: 14) acknowledgement that 'the theory of performativity is not far from the theory of hegemony [...]: both emphasize the way in which the social world is made [...] through a collaborative relation with power'. The quote highlights the close relation between performativity and hegemony as both consider power central to any social and political organization. Mouffe (2005b: 17) agrees with Butler in this respect, considering 'the concept of hegemony [as] the key notion for addressing the question of "the political"'. In the following, I look at the genealogy of spaces of politics through these two theoretical lenses, suggesting that they fruitfully complement each other.

Gregson and Rose (2000) draw on the concept of performativity as Judith Butler developed it. Hence, before taking their notion of performativity 'elsewhere' – namely to political geography – it is worthwhile recalling Butler's notion

---

8   Against a tendency in feminist geography to focus on political spaces and by doing so expanding the boundaries of a narrow, masculinist view on the politics as electoral politics (Brown and Staeheli 2003, England 2003, Kofman 2008, Staeheli 1996), I would like to take a feminist view on spaces of politics. As spaces of electoral politics mirror the power relations inherent in society, I consider the study of institutionalized politics as a possibility to trace both the genealogies of these power relations that constitute spaces of politics and the contestation of these power structures.

of performativity. In the new preface for a second edition of Gender Trouble, Butler (1999 [1990]: xv) writes:

> 'I originally took my clue on how to read the performativity of gender from Jacques Derrida's reading of Kafka's "Before the Law". There the one who waits for the law, sits before the door of the law, attributes a certain force to the law for which one waits. The anticipation of an authoritative disclosure of meaning is the means by which that authority is attributed and installed: the anticipation conjures its object. (...) In the first instance, then, the performativity of gender revolves around this metalepsis, the way in which the anticipation of a gendered essence produces that which it posits as outside itself. Secondly, performativity is not a singular act, but a repetition and a ritual, which achieves its effects through its naturalization...'

In the quote, Butler understands performativity as an expectation that ends up producing the very phenomenon that it anticipates. While Butler has extensively shown what this expectation means for the way gender is sexually regulated, her reference to Kafka's (1998 [1925]) 'Before the Law' is interesting when framing spaces of politics as performative. In Kafka's parable, the Law is figured in a spatial way as a room with a door, which a man from the country wants to enter. This place where the Law is supposed to exist is believed to be open for all but is actually guarded by a series of doors and doorkeepers. The space of the Law is pictured as a room with a door in the parable. In liberal democracies, the 'Law' is spatialized through the materialization of the law in institutions like the court. In a performative sense, however, the edifice of the courthouse does not pre-exist the performance or the deed of judging, hearings, and passing laws. It is these performances that bring the space of the Law, the courthouse, into being. These spaces of the Law can be read as one example for spaces of politics, for spaces where politics are 'done'.

Hence, spaces of politics are performative in the same way as spaces of the Law as they are also brought into being by reiterative and citational practices that produce the effect they name. Along with Kuus (2007: 91) who argues for refocusing research in political geography away from subjects of identity and toward the practices through which subjects are made, I plead to refocus on the citational practices that bring the spaces of politics into being. Performative spaces of politics do not emerge from a singular political act, but from a reiteration of norms that have assumed their political status through their repetition. Hence, spaces of politics can be re-figured as 'imitations' (Butler 1993: 125) with no original. By thinking spaces of politics as iterative imitations with no original, the Eurocentrism of politics is pushed from its pedestal of providing the origin of political organizing (see Campbell and Harbord 1999: 230 for the analogy to Butler's work). A performative approach to politics offers the possibility to trace the genealogy of politics, the history of the way spaces of politics were imagined and brought into being through discursive practices, without falling into the Eurocentric trap of considering the Greek polis – and more generally Western democracies – as the origin of politics and hence as normative archetype. Hence, performativity offers the possibility to 'de-naturalize' (Butler 1999 [1990]: xxi) or rather 'de-colonize' the apparent natural colonial order of political spatialities. In the same way as Butler questions the history of sex, it can be asked: Do politics and their spatialities have a history? Are the norms and rules associated with the political discursively produced by (scientific, geopolitical) discourses in the service of (geo-)political and social interests?

The genealogy of the political has been traced in a de-naturalizing and de-colonizing way by Carole Pateman and Charles Mills in their respective books 'The Sexual Contract' (Pateman 1988), 'The Racial Contract' (Mills 1997) and in 'Contract and Domination' (Pateman and Mills 2007). Their work is crucial when trying to understand spaces of politics as hegemonic institutions and reveal the postcolonial history of these spaces. Pateman (1988: 1) writes in the opening of her book:

> 'The most famous and influential political story of modern times is found in the writing of the social contract theorists. The story, or conjectural history, tells how a new civil society and a new form of political rights is created through an original contract. An explanation for the binding authority of the state and civil law and for the legitimacy of modern civil government is to be found by treating our society as if it had originated in a contract'.

Pateman and Mills critique contract theory as Thomas Hobbes, John Locke, Jean-Jacques Rosseau, Immanuel Kant and John Rawls have developed it for missing half of the story, which tells how modern forms of patriarchy, racism, and imperialism are established. They show how imaginaries of the political inspired by classic contract theorists have justified the patriarchal, racial, and imperial structures that have shaped the political order and the very spaces of politics[9]. Pateman (1988: 220–221) emphasizes that *white men* have sealed the original contract. Through this act, white women and nonwhites were performatively constructed as second-class citizens or not as citizens at all. White women and nonwhites were positioned outside the political and were denied access to the spaces of politics – just like the man from the country in Kafka's parable. Hence, the spaces of politics are performatively brought into being through the juridical regimes of the racial and sexual contract. The racial and sexual contracts are examples how in a performative analysis of the political 'there is no power, construed as subject, that acts, but only (...) a reiterated acting that *is* power' (Butler 1993: 171), in this case acting consists of a reiterative citation of the contract. Tully (2000: 44) calls attention to the way in which much contemporary political theory obliterates any discussion of embarrassing origins and condemns political theory for claiming that contract theory is considered 'an abstract starting point [...] that had nothing to do

---

9   It is important to highlight that Pateman and Mills discuss the similarities and differences between gender and race. While they consider 'masculinity', 'femininity', and 'race' as political constructs that result from the language of nature employed in contract theory, they argue that gender (for the division between childbearing and non-childbearing halves of humankind) has a much longer history than race, which only comes into existence in the modern period (Pateman and Mills 2007: 5).

with the way these societies were founded'. To read his statement against the backdrop of performativity means that the political spaces of contemporary democratic societies are produced by the recitative power of discourses of political (contract) theory that was employed to establish patriarchal and imperial (geo-) politics.

The way spaces of politics are performatively brought into being through patriarchal and imperial imaginaries of the political can be traced and made visible along the example of postcolonial democracies. The case of Ecuador – or the Andes more generally – highlights the power of these Eurocentric and imperial discourses that served to normalize and naturalize the white man as the hegemonic political subject even after formal independence.

With the Spanish conquest a *new world* appeared on the horizon of European imaginaries, being considered as

> 'the tabula rasa on which the principles and accomplishments of Western rationality (religious beliefs, scientific advances, and humanistic paradigms [including political order]) could and should be inscribed' (Moraña, Dussel, and Jáuregui 2008: 7).

The white men's superiority was naturalized by dividing the colonial political order into two parallel 'republics' (the *República de los Españoles* and the *República de Inidos*), relegating indigenous people outside the colonial spaces of politics. Further, heterosexuality, referred to by Young (1995: 25) as the 'implicit politics of heterosexuality', and 'gender relations have provided a template for the organization of relations of power and difference' (Radcliffe 2000: 172). Material laws, property laws and land laws gendered the political architecture of imperialism (McClintock 1995) in a way that women – white, indigenous and Creole women to different extent – were positioned by these juridical regimes outside, or in Kafka's word 'before', the spaces of politics as the colonial regime denied them access to political participation and citizenship. The colonial marking of spaces of politics as white and masculine can be seen as a result of the citational practices like the indigenous tribute and the constant effort to (re-)produce the gendered and racialized political ideas, routines and practices from the metropolis in the colonies. As the performances do not outlast the moment of their acting, the colonizing acts had to be repeated permanently in order to reassert the (post-)colonial gendered and racialized order. Both gendered and racialized power dynamics were from the outset fundamental to the securing and maintenance of colonial discourse and determined also the constitution of politics in post-colonial times. With independence from Spain, liberal ideals based on contract theory were (re-)produced to legitimate the privileges of Creoles (American-born whites) while at the same time 'retaining aspects of Spanish colonial legislation and institutions ensured the continued subjugation of Indians' (Clark and Becker 2007: 8). The power of the colonial juridical regimes becomes visible in the reproduction of these racialized and gendered structures after the termination of Spanish rule, e.g. in the perpetuation of the indigenous tribute (Prieto 2004) or the denial of suffrage for women, illiterates and people without property (Dore and Molyneux 2000, Prieto and Goetschel 2008, Radcliffe 2002, Schurr 2009b). The effects of the colonial logic of ordering and organizing spaces of politics still reverberate in the post-colonial spaces of politics in form of the colonial architecture of municipalities, imaginaries of authority, political rituals and proceedings etc. as the colonial painting of three Spanish governors that decorates the municipality in highland Chimborazo (Figure 8) and the interview quote from a female indigenous mayor show:

Figure 8: Colonial painting in the municipality of Riobamba, Chimborazo, Source: Author

'They *(el pueblo)* stigmatized me first because I am a woman and second, for the color of my skin. For them, a mayor had to have blue eyes, blond hair and a European stature...' (Cecilla Mantilla, mayor, 2009).

Through this kind of practices and imaginaries the hegemonic 'nature' of political order along gendered and ethnic cleavages was brought into being and constantly (re-)produced. Hence, to trace the history of spaces of politics requires – in Mouffe's words – to recognize

'the hegemonic nature of every kind of social order and the fact that every society is the product of a series of practices attempting to establish order in a context of contingency' (Mouffe 2005b: 17).

The hegemony of any political order is brought into being through practices of antagonism between different political identities. Mouffe and Laclau (1985: xvii) emphasize that political identities are not pre-given but constituted and reconstituted through debate in the public sphere. Thomassen (2005: 638) adopts Mouffe and Laclau's notion of political identities as hegemonic to the very spatiality of politics, stating that 'political space and time, too, are hegemonically constituted'.

While Mouffe's conceptualization of politics as hegemonic is crucial to analyze power relations and resulting in- and exclusions that constitute the spaces of politics, it is Butler's notion of performativity that helps to capture these practices that establish the hegemonic order in the first place. The relation between hegemony and performativity in the constitutional process of political spaces is pointed out more explicitly in Laclau's (1996: 90) remark:

'the two central features of a hegemonic intervention are, in this sense, the "contingent" character of the hegemonic articulations and their "constitutive" character, in the sense that they institute social relations in a primary sense, not depending on any a priori social rationality'.

The second feature is crucial for the endeavor to conceptualize spaces of politics as performative oriented towards Gregson and Rose's performative spaces: the hegemonic political articulations (in form of constitutions, suffrage, laws) constitute the spaces of politics through sedimented practices that result from and at the same time consolidate, strengthen and reinforce the hegemonic political order – the spaces of politics do not exist a priori of these hegemonic articulations. The first feature, the 'contingency' of hegemonic (spatial) articulations is central within the concept of antagonist politics and serves as a bridge to the next section that deals with the contestation of a hegemonic political order in form of counter-hegemonic resistance and subversion. According to Mouffe (2005b: 18), every (political) order is a temporary and precarious articulation of contingent practices – 'things could always be otherwise'. Any temporary and spatial order is 'political' since it is the expression of a particular structure of power relations. The colonial political order in the Andes e.g. was a temporary and precarious order that constantly needed to be reproduced to maintain hegemonic power relations. The apparently 'natural' Spanish hegemony was actually the result of sedimented practices. The fact that any hegemonic order is based on some form of exclusion – in case of the colonial political order the exclusion of indigenous people, *mestizos*, Afro-descendants, and women and their imaginaries of political ordering – means that 'there are always other possibilities that have been repressed and that can be reactivated' (Mouffe 2005b: 18). Hence,

> 'every hegemonic order is susceptible of being challenged by counter-hegemonic practices, i.e. practices which will attempt to disarticulate the existing order so as to install another form of hegemony' (ibid.).

Mouffe's thinking about the contingency of the political order resembles Butler's endeavor to question contemporary – or in Mouffe's words hegemonic – notions of reality and think about the way subversive – or counter-hegemonic – practices institute 'new modes of reality' (Butler 2004b: 217). How the hegemonic order of spaces of politics becomes contested and challenged by counter-hegemonic practices and how possible 'new modes of reality' look like in the case of Ecuadorian politics will be addressed in the following section.

CONTESTING (POST-)COLONIAL POLITICAL HEGEMONIES

> 'The political arises when the given order of things is questioned; when those whose voice is only recognized as noise by the policy order claim their right to speak, acquire speech, and produce the spatiality that permits and sustains this right' (Swyngedouw 2008: 24).

As highlighted, Mouffe as well as Butler and Gregson/Rose are interested in questions of how social change occurs within and despite of a hegemonic order. All of them depart from a poststructuralist epistemology that denies any pre-discursive essence and focuses rather on the contingency, ambiguity, lack, slippage that constitute discourses, identities and spatialities. Hegemonic orders – be it a gender or a democratic order – are never stable in their view because of the ambiguity, contingency or lack in any discursive formation of identity and spatialities. The way, however, they picture social and political change differs slightly. While both Mouffe and Butler assume that change occurs when hegemonic power relations are challenged by counter-hegemonic practices (Butler 1997: 160, Mouffe 2005b: 33), they employ a different vocabulary to conceptualize processes of change. Butler speaks of subversion and slippage, whereas Mouffe refers to processes of disarticulation-rearticulation that constitute an antagonistic politics. In my view, the main difference between their approaches is the type of political agency inscribed in the concept of subversion and disarticulation respectively. For Mouffe (2005a: 12, 2005b: 19), one of the main tasks for democratic politics consists in defusing the potential antagonism between collective political identities of 'we/they' that exists in any political order. In 'Hegemony and Socialist Strategy', Laclau and Mouffe (1985: 170–171) highlight the importance of collective political identities such as the 'new social movements' for an antagonistic politics, even though they emphasize that these collective identities are constructed and hence are subject to an always fluid and instable character along floating demarcating lines. In contrast, Butler's call for 'repetitious citation of subversive interventions rejects the traditional model of mass politics' (Chambers and Carver 2008: 46). Butler does not put forward a theory of subversion and provides little explicit conceptualization of the term. She argues that the norms that govern what is (not) real and who is intelligible as a (political) subject, are called into question and reiterated at the moment in which performativity begins its citational practice. In other words, because citational performance needs to be reiterative to reproduce its effect, slippage is not only possible, but inherent to the very process of reiteration. Hence, 'it is in the slippage inherent to discourse, that both intentional and unintentional change occurs' (Haller 2003: 763). Although Butler rejects models of mass politics, her conceptualization of performativity as sedimented practices requires not only a single act of resistance, but also the reiteration of these acts. The constant repetition of subversive acts – like drag – is necessary to destabilize apparently 'natural' or 'hegemonic' – gender or political – orders by showing that these orders are not 'natural' but rather 'follow from contingent, historical, malleable practices that could have been otherwise, and could be different in future' (Chambers and Carver 2008: 23). In fact, Butler's aim to destabilize the power through which the troubling norms of gender are naturalized is not far from Mouffe's assumption that any hegemonic order is susceptible to being challenged by counter-hegemonic practices. Both assume that counter-hegemonic practices are able to challenge a hegemonic order, be it in the case of Butler of heteronormativity or as in the case of Mouffe of democracy.

For Mouffe (2005b: 33), it is the agonistic struggle that 'should bring new meanings' through 'a process of disarticulation of existing practices and creation of new discourses and institutions'. Drawing on Gramsci's notion of disarticulation, Mouffe (1979: 193–194) locates transformation within the political struggles between two hegemonic principles to appropriate ideological elements in a process of *disarticulation-rearticulation* of given ideological elements. Emphasizing the 'composite, heterogeneous, open and ultimately indeterminate character of democratic tradition' (Mouffe 1993: 17), she highlights that because a political order at a given moment is the result of relations of forces, it is the object of a perpetual process of transformation. In other words, transformation can be realized because a hegemonic order is only a precarious articulation of contingent practices that is always already contested through the possibility that things could be otherwise or, more precisely, that a political order could be imagined and later on rearticulated in a different way. Especially Laclau has focused in his work on the transformative potential of social movements, arguing that political change occurs when counter-hegemonic subjects articulate their 'demands'. Any hegemonic order consists of a confrontation between unfulfilled demands on the one hand and an unresponsive power on the other hand (Laclau 2005: 86). At the beginning, the series of isolated demands articulated by diverse movements have nothing in common than demanding something from the political order and authorities. The only common characteristic of these diverse demands is the fact that they are opposed to the hegemonic system that they consider as lacking, repressive or menacing. It is the experience of lack, repression and menace that first leads to general frustration of a series of social demands before it facilitates the movement from isolated democratic demands of particular movements to more popular ones, articulated in a collective counter-hegemonic subjectivity.

In the following, I would like to exemplify along the case of Ecuadorian politics, how apparently 'natural' colonial and post-colonial orders are questioned by counter-hegemonic subjects, showing that every order brings along new forms of exclusion that result in new antagonistic struggles and imaginaries of the political. Focusing on different ways spaces of politics are brought into being through everyday practices, I show how through disarticulation, resignification and slippages the very spatiality of politics is contested and transformed.

PERFORMING ANTAGONISM IN CHIMBORAZO'S LOCAL POLITICS

The province of Chimborazo has the highest concentration of people self-identifying as indigenous in Ecuador (approximately 50 percent). Indigenous people in Chimborazo were among the first to question the hegemonic (post-)colonial order in Ecuador, when in 1871 they protested against the indigenous tribute, a tax demanded by the church and the state (Clark and Becker 2007: 11). From then on, indigenous people fought against the state, demanding more autonomy, less economic and social exploitation, and land reforms. Their 'demands' (Laclau 2005) were clearly opposed to the hegemonic system they experienced as repressive and menacing. In 1974, Lázaro Condo, an indigenous protester, died in the midst of a confused confrontation between the police and indigenous members of the *comuna* Tocetzinín in Chimborazo who occupied a disputed plot of land. Becker (2008a: 144) states that his assassination coincided with 'a shift in the content, discourse, and strategies of indigenous movements'. This shift consists in Laclau's terms of a collectivization of particular demands as the indigenous protest against the hegemonic order was increasingly joined by civil rights, students, women's and popular movements that questioned the status quo. It is the increasing collective opposition of these different movements that results in a new political configuration which triggers 'change in Indian ethnic identification' (Crespi 1981: 478) and their relations with the whitened spaces of politics.

The increasing antagonism between the white/*mestizo* state and indigenous movements culminated in 1990 in a nationwide *levantamiento* (uprising) consisting of massive blockades that shut down the country for a week (see Introduction). The political space of the indigenous uprising was brought into being through mass mobilizations citing popular forms of protests by counter-hegemonic political collectives in Latin America. The protests transformed streets and plazas into political spaces through the presence of their counter-hegemonic bodies and demands towards the state. The protests forced the government to negotiate the demands of the indigenous protesters. In the negotiations, the boundary between the counter-hegemonic political spaces and the hegemonic spaces of politics was blurred. To speak in Kafka's words, the (indigenous) man from the country was permitted access to the hegemonic spaces of politics – in this case the edifice of the National Congress. The clear-cut boundary between indigenized political spaces and white/*mestizo* spaces of politics became even fuzzier when in 1996 the indigenous and peasant party *Pachakutik* was founded and indigenous movement politics became institutionalized within the very spaces of politics (see Chapter 1).

In Chimborazo, the municipality of Guamote was among the first municipalities headed by an indigenous mayor. Mariano Curicama's election as mayor – concurrent with the election of two indigenous municipal councilors – reflected a rapid shift in political power away from the traditional, white and *mestizo* feudal elites. In 1991, *mestizo* elites, in fact, tried to prevent Mariano Curicama from taking office. Indigenous organizations responded with a threat to boycott *mestizo* businesses. This threat cleared his path, together with the accompaniment of 5000 indigenous supporters, who literally stood behind him as he entered the building of the municipality (Van Cott 2008: 155). This incidence shows not only the often forceful antagonism that accompanies processes of change, but the centrality of spatial appropriation. In order to 'become' mayor, Mariano Curicama had to enter the building of the municipality. It was through the bodily appropriation of the municipality that he actually challenged the hegemonic spatial order of politics. The sheer presence of bodies that are racialized as indigenous through colonial discourses within spaces of politics that have been marked as white for centuries destabilizes the hegemony of the post-colonial political order. It was the sedimentation of counter-hegemonic practices such as protests, political organization of indigenous communities, forming of alliances with other social movements and the Catholic Church over decades, that finally resulted in the contestation of the hegemonic (post-)colonial order of spaces of politics – including 'the foundations and contours of contemporary democratic practices' (Yashar 1999: 76). Through the iteration of these subversive practices, the norm that defined who is considered an intelligible political subject was challenged. The norms have been expanded on the one hand by abolishing the literacy requirements for citizenship in the constitution of 1979 and on the other hand through processes of cultural change.

One central achievement of the indigenous movement was the resignification of the term 'indigenous' itself. Since colonial times, addressing someone as '*indio*' not only '*reflected*' a relation of social domination; but actually '*enacted*' domination (Butler 1997: 18)[10]. By taking up the term '*indio*' in an affirmative way by the indigenous movement, a reverse interpellation (Butler 1993: 83) was effected that resulted in the generation of a subjectivity of resistance and resignified the very term '*indio*'. The politicization of the indigenous movement, however, not only challenged the intelligibility of political subjectivity, but also the 'political organization and spatial orders' (Radcliffe, Laurie, and Andolina 2002: 291) of Ecuador's post-colonial democracy.

The provincial council of Chimborazo and the municipality of Riobamba (the capital of the province) serve as example to look at the way spaces of politics have (not) been transformed under indigenous authority[11]. Nowadays, Mariano Curicama is – in the second term – the head of the province of Chimborazo, the prefect. It is not only the presence of his racialized body and that of many indigenous people who wait to be attended by Curicama that indigenizes and hence decolonizes the spaces of the provincial council, but the bilingual signage in Spanish and *kichwa* within the building that helps citizens to find their way around, the indigenous *whipala* flag that decorates the floors of the building, and the announcement of a number of activities like the invitation for a participatory assembly to define participatory budgets or a song contest for indigenous women's groups at the 8th of March that traditionally had not taken place within the building of the provincial council. While the appropriation of this (post-)colonially whitened space of politics takes place through resignifying a new meaning to the spatiality to which indigenous people frequently had been denied access during (post-)colonial times, the very foundation of democracy is challenged by participatory modes of politics. Further, the frequent visits of Curicama in the rural indigenous communities temporarily erect spaces of politics in the country, decentering the urban monopoly as site for institutionalized politics.

---

10 Radcliffe (2000) highlights that in the context of nation-building, indigenous identities are unintelligible apart from their struggle with the state, as the nation-state attempts to shape 'indigenous-ness' within its remit of forging the national community, while indigenous groups infuse representations and discourses with elements of their own making. However, as the state project of dealing with indigenous groups has changed over time, depending on the prevailing political and economic agendas, so too has the nature of 'indigenous-ness' changed.

11 The following analysis is based on fieldwork conducted in the province of Chimborazo between 2008 and 2010, including ethnographic observation, interviews with politicians and representatives of social movements (for a broader reflection on the methodology see Introduction and chapter 4 and 8).

From this evidence, one could assume that the antagonistic struggle between hegemonic (white/*mestizo*) subjectivities and counter-hegemonic (indigenous) subjectivities and imaginaries of political order has resulted in an initiating process of decolonization. It seems that a new, more inclusionary, pluralist democracy has been established that includes formerly excluded social groups. As Radcliffe et al. (2002: 299) highlight, however, 'alongside the expansion of political opportunity lie exclusionary patterns of political culture that limit participation'. While the antagonistic struggle of the indigenous movement against the white and *mestizo* political elites permitted the insertion of indigenous citizens into spaces of politics, new antagonisms and exclusions persist in making that access unequal and/or denying the access at all to certain members of subaltern groups. In the province of Chimborazo for example, the gendered political culture of indigenous organizations and the 'cultural constructions of non-authoritative and apolitical femininity' (Radcliffe, Laurie, and Andolina 2002: 300) still position women at the margin of spaces of politics just like the man from the country in Kafka's parable. By doing so, indigenous political orders (re-)produce the apparently 'natural' masculinity of spaces of politics, as the statement by one of the few female indigenous politician evidences:

> 'It is that ... always politics, all the life, politics have always been done by men – everything – be it organizations of the [indigenous] *comunidades*, [...] be it in the political parties, even in *Pachakutik* everything is ruled by men' (interview female, indigenous town councilor, 23.03.2010, Colta).

Indigenous women in the Province of Chimborazo increasingly start to organize themselves in form of the *Red de Organizaciones de Mujeres Indígenas de Chimborazo*. The network also offers workshops – financed mainly by international development organizations – that aim to organize, train and inform the women with regard to their political rights and participation with some success as the following quote from a local indigenous female politician demonstrates:

> 'Before [the gender quota law], we did not know anything, they [the men] told us to vote for someone and we all voted for this person. Sure, there have always been a few women, but we, the women themselves, didn't vote for the female candidates, because we believed that men are the only ones who can do politics' (interview female, indigenous president of a local government, 03.04.2009, Riobamba).

Despite the mentioned gender quota law, introduced in 1998, however, only few female candidates are actually elected in the local elections (CONAMU 2002). Especially indigenous women with long experience in 'informal' politics more and more question the masculine hegemony inherent in indigenous spaces of politics. Hence, a new antagonism emerges between indigenous women and indigenous politicians like Curicama who in the meantime have turned into hegemonic political subjects – at least in the province of Chimborazo.

New forms of antagonism not only occur along gendered cleavages, but also along ethnic, urban/rural, and party lines. While Mariano Curicama is respected within his own racial group, the indigenous prefect faces *mestizo* recalcitrance in the ethnically divided province. Due to the colonial ordering of space, the *mestizo* population is mostly located in the urban centers of the province until today. As a consequence, the antagonism between indigenized and *mestizo* spaces of politics is materialized along an urban/rural jurisdiction of the respective political institutions. While the provincial council represents the rural areas of the province and is in charge of the development of mainly indigenous (rural) communities, the municipality historically represents the *mestizo* urban center. This (post-)colonial political spatial order is reflected in the respective spaces of politics: in the municipality no sign of indigenous spatial appropriation is visible. The colonial building of the municipality is decorated with colonial paintings (see Figure 8), the mayor is a *mestizo* man and few indigenous people can be spotted within the municipality. The hegemony of a *mestizo* political culture became evident to me, when I witnessed how an indigenous woman started to speak in *kichwa* to the mostly *mestizo* politicians in a town council meeting. People got very nervous, the secretary asked how she was supposed to protocol the request of the woman (ethnographic note, 15.2.2010). Probably, Judith Butler would have been enthusiastic about the first evidence of a slippage in a hegemonic *mestizo* space where indigenous people, dress and language were excluded since colonialism. Why hadn't there been any citizens before who spoke out their claims in *kichwa* in a town council meeting? Certainly, it did not occur to any citizen to speak in *kichwa* as the ones they addressed would not understand them. Since the last election, however, there has been one indigenous woman in the town council and she was the one the *kichwa* speaking woman addressed in her speech. The fact that an indigenous town councilor was present needs to be seen as an effect of the performative politics of language. Thanks to indigenous pressure in the constitutional negotiation a new political representative was introduced at municipality level: the rural town councilor. While this function seems a paradox in itself, it's the colonial ordering of political territoriality that explains this position. The – mainly indigenous – rural communities that are part of the municipality's territory elect the rural town councilor. As the rural areas have been (post-)colonially neglected by – mainly *mestizo* dominated – municipal policies, the establishment of rural town councilors in 2008 can be considered as result of the demands of the indigenous movement for political representation. Hence, the incidence of the first *kichwa* dialogue in the town council shows the material effects produced by performative utterances inscribed in the constitution. Shifts in the political language towards decolonization, including a better representational system of rural areas, produce a decolonized space of politics in which an indigenous woman is for the first time able to speak in her mother tongue.

In sum, in the province of Chimborazo an antagonism takes place between two differently racialized spaces of politics: the municipality and the provincial council. While the municipality is constituted by the hegemony of *mestizo* political subjectivities and practices, in the provincial council former counter-hegemonic subjects have turned into the hegemonic subjects that define the political order and agendas. Both spaces are highly exclusionary as indigenous people have very little voice in the municipality and *mestizo* people feel neglected by the policies of indigenous Prefect Curicama whose 'participative' budgets benefits mostly indigenous communities (personal communication with a functionary of the provincial council, 20.3.2010). The spaces of politics like the municipality or the provincial council in the province of Chimborazo are object and product of ongoing antagonistic struggles between different collective identities that are constituted along specific demands. To say that antagonism occurs only along gender and/or ethnic lines, however, would simplify the diverse antagonistic relations that constitute the political spaces of the province. Antagonism further can be observed, for example, within the indigenous movement between mostly Catholic supporters of *Pachakutik* and *Amaute*, an indigenous party 'sponsored' by the evangelical church that gains more and more support in indigenous communities (Van Cott 2008: 153), with regard to the question whether indigenous organizations would be better off not to mingle with electoral politics, and between poor *mestizo* and indigenous peasants in their fight for financial support of the state and international development agencies.

The empirical case study aimed to highlight how antagonisms are spatially materialized. The 'spatial contradictions' (Lefebvre 1991: 365) between the indigenized space of the provincial council and the *mestizo* space of the municipality are a result of and at the same time express a particular structure of power relations. Both spaces were brought into being through the sedimentation of certain (racialized and gendered) practices that are enabled and disciplined by certain discourses. While the practices and imaginaries that constitute the municipality are disciplined by (post-)colonial understandings of political order, indigenized ways of doing politics challenge the hegemony of this order by framing their practices within discourses of decolonization, resistance, and alternative development. Following the antagonistic logic of resistance studies (Rose 2002) that aim to highlight how hegemonic spaces of politics are subverted, appropriated and contested, the story could end at this point, as it has been shown how the (post-) colonial white and *mestizo* 'nature' of spaces of politics has been successfully challenged through social movement protest and how spaces of politics have been appropriated by counter-hegemonic subjects. Along with Mouffe, however, I would like to go one step further by asking how the kind of spatial antagonism encountered in the province of Chimborazo could be transformed into agonism with the aim to facilitate a pluralist democracy. In other words, how could antagonistic struggles be 'tamed thanks to the establishment of institutions and practices through which the potential of antagonism can be played out in an agonistic way' (Mouffe 2005b: 20–21)?

INTERCULTURALIDAD AS AGONISTIC POLITICS

'What resources must we have in order to bring into the human community those humans who have not been considered part of the recognizably human? That is the task of a radical democratic theory and practice that seeks to extend the norms that sustain viable life to previously disenfranchised communities' (Butler 2004: 225).

'Coming to terms with the hegemonic nature of social relations and identities, it [an agonistic approach] can contribute to subverting the ever-present temptation existing in democratic societies to naturalize its frontiers and essentialize its identities' (Mouffe 2005b: 105).

Departing from Butler's notion of performativity – especially developed in her early work 'Gender Trouble' (1990) and 'Bodies that Matter' (1993) –, Gregson and Rose show along their empirical research projects on car-boot sales and community art that 'space is citational, and itself iterative, unstable, performative' (Gregson and Rose 2000: 447). In the previous sections, I have outlined how a performative approach to spaces of politics reveals the power relations and related effects of naturalization of a hegemonic order that constitute the spatialities of politics. The concept of performative space as suggested by Gregson and Rose certainly is a helpful tool for a critical political geography to conceptualize processes of hegemonialization (Mouffe), naturalization (Butler), counterhegemonialization (Mouffe) and subversion (Butler) that bring the spatialities of politics into being. In the following, however, I argue that in order to fully realize the potential of Butler's work for political geography, it is crucial to draw on her more recent work (Butler 2004a, 2004b, 2009, Butler and Spivak 2007) in which she engages more directly with questions of social transformation, norms, politics, and democracy. Butler's recent work does not focus on performativity as explicitly as her earlier work. Performativity as a tool to denaturalize established norms stays crucial, however, in her call to 'expand our fundamental categories' to make them 'more inclusive and responsive to the full range of cultural populations' (Butler 2004b: 223). It is in the chapter 'The question of social transformation' that she links this call to the theory of radical democracy as it was outlined by Mouffe and Laclau (1985). She considers the extension of norms that 'sustain viable life to previously disenfranchised communities' as a central 'task of a radical democratic theory and practice' (Butler 2004b: 225). By doing so, she emphasizes that norms define the intelligibility of a (political) subject and that as long as someone is not recognized as intelligible one cannot enter in an antagonistic – or even oppressed – relation with the hegemonic subjects. Mouffe, in contrast, does not pay attention to the question of intelligibility, but departs from the assumption that the political is constituted by an antagonistic relation between intelligible subjects who are identifying with binary collective identities of us/them. For her, the most important challenge for contemporary democracies consists in establishing 'this us/them discrimination in a way that is compatible with pluralist democracy' (Mouffe 2005a: 101). Butler and Mouffe come together in their search to 'generate new possibilities' (Butler 2004b: 194) of conviviality within a pluralist democracy, even so they suggest different modes to achieve this goal: the expansion of the norms of intelligibility (Butler) and the transformation from antagonism into agonism (Mouffe). In other words, both theorists are centrally concerned in thinking through a way to create a social and political order that embraces difference, challenges norms (and practices of disciplinary normalization) and overcomes mechanisms of exclusions, violence and oppression.

While Butler (2004b: 225) advocates a process of cultural translation to achieve this aim, Mouffe (2005a: 98–105, 2005b: 20–21) proposes an agonistic model of democracy which channels dissenting voices in a non-violent way. For Mouffe, the aim of democratic politics is to transform antagonistic relation between enemies into an agonistic relationship in which the conflicting parties recognize the legitimacy of their adversary. Because they are adversaries and not enemies, 'they see themselves as belonging to the same political association, as sharing a common symbolic space within which the conflict takes place' (Mouffe 2005b: 20). Mouffe's model assumes that any society is structured around the agonistic configuration of power relations 'between opposing hegemonic projects which can never be reconciled rationally' (Mouffe 2005b: 21). It is the task of a pluralist democracy to create a set of democratic institutions and procedures that regulate these agonistic confrontations between the two adversaries that both accept these institutions and procedures.

Mouffe does not spell out empirically what kind of ideologies, institutions, and everyday practices are able to channel dissenting voices in a radical democracy. In the following, I ask to what extent the concepts of *interculturalidad* and *plurinacionalidad*[12] that have been introduced to Ecuador's constitution in 1998 can be seen as political utopias in Mouffe's sense of agonism. The concept of performativity serves once again as a crucial tool to grasp the processes of spatial materialization that challenge the white/*mestizo* hegemony by imagining and performing spaces of politics as intercultural. *Interculturalidad* as a concept – in a similar way as agonism – departs from the configuration of power relations around which a society – in this case the post-colonial society of Ecuador – is structured (in contrast to multiculturalism that seeks to harmonize conflictive relations between different groups). The conflictivity that results from the historical power relations between hegemonic and counter-hegemonic groups in a post-colonial society is inherent in and made visible through the concept of *interculturalidad* as it recognizes the existence of divergent epistemologies, ideologies and interests. Walsh (2003: 112) highlights that *interculturalidad* is understood as both a process and project to confront and transform power relations and the hegemony of political institutions that excludes certain social groups. One could say that *interculturalidad* is the postcolonial version of agonism as it

---

12 While pluriculturalism (convivality of different cultures in a post-colonial context) and multiculturalism (inclusionary occidental model of a neoliberal state) are descriptive terms that indicate the existence of different nations/cultures in a specific territory, *interculturalidad* still does not exist but is a project to be constructed. *Interculturalidad* means that there are two distinct cosmologies at work, a Western and an Indigenous one. *Interculturalidad* aims to construct a new society on the basis of a dialogue between these two epistemologies over the erosions of the liberal and republican state (Walsh 2009).

'proposes a model of political organization for decolonization aimed at recovering, strengthening and democratizing the state [...], transforming the structures and institutions in order to recognize political and cultural diversity' (Walsh 2009: 78).

Since the 1990s, *interculturalidad* has been a crucial demand by both indigenous and Afro-Ecuadorian movements that was first addressed by the state in the 1998 Constituent Assembly (Walsh 2008b: 507). Through a performative speech act, Ecuador was declared a 'plurinational and multiethnic state' (Art 1.) by the constitution of 1998. The constitution of 2008 slightly altered the speech act when stating in its first article that 'Ecuador is an intercultural and plurinational state'. The sedimentation of these speech acts has been realized through their iterative citation in political speeches, political agendas (Figure 9), and academia. The ongoing transformation of the political order towards an intercultural space of politics, however, not only has taken place through a discursive reframing, but also through everyday practices. The presence of indigenous and Afro-Ecuadorian people in important political positions (Figure 10), their declaration as nationalities, and their cosmologies, e.g. the concept of *sumak kawsay*, an indigenous philosophy of 'living well' (see Radcliffe 2011), challenge the uni-national and mono-cultural framework of Ecuador's post-colonial democracy. Especially in local spaces of politics, serious attempts have been made to translate the discourse of *interculturalidad* into political practices by creating for example intercultural round tables (*mesas interculturales*), introducing participatory budgeting processes that take into account the divergent interests and necessities of different social groups, conducting workshops and discussions around alternative visions of (local) development and participation (own ethnographic notes). These political progressive initiatives aim to re-think and re-found spaces of politics interculturally by encouraging 'politics of convergence, of conviviality' (Walsh 2009: 71) between ideologically, epistemologically, ethnically different collective identities.

Figure 9: Cover of the political agenda of CONAIE
Source: CONAIE Homepage

Figure 10: Indigenous leaders in the National Assembly
Source: Author

The post-colonial white/*mestizo* hegemony that has dominated spaces of politics is challenged through the sedimentation of these intercultural practices. Intercultural imaginaries propose a mode of political order that recognizes difference of ideologies and epistemologies and provide a space, e.g. in the social space of the intercultural round table, where agonistic struggle between different groups can literally take place. While these spaces exist and the discourse of *interculturalidad* has gained importance within the discursive political field of Ecuador, a critical evaluation of the diverse local spaces of politics that have been labeled 'intercultural' shows that new forms of antagonism rather than agonism constitute these spaces. Lalander (2010: 506) refers to the antagonism that results from intercultural politics as 'intercultural dilemma'. It emerges if an indigenous or Afro-Ecuadorian mayor starts providing welfare for all social sectors, without prioritizing indigenous or Afro-Ecuadorian grievances. The protest of these social groups certainly needs to be seen as reflection of centuries of exclusion and frustrated expectations, but to re-found spaces of politics as intercultural would demand to respect the *mestizo* population as legitimate adversaries with whom an agonistic confrontation is possible over the distribution of resources. Hence, the political reality evidences that '*interculturalidad* still does not exist, but is a project to be constructed' (Walsh 2008a: 140). With this permanent lack, *interculturalidad* presents an example of a social imaginary or horizon as discussed by Laclau (1990: 64), that is, as a constant social demand that due to the dialectical relation between hegemony and lack, however, can never be complete(d).

Interculturalidad can be considered as a political project and practice in post-colonial societies that could actually facilitate an agonistic radical democracy. Hence, intercultural spaces of politics can be imagined as spaces that include marginalized, oppressed and unintelligible subjects, provide room for dialogue between different ideologies and epistemologies and overcome exclusionary and antagonistic modes of political order. These decolonized, intercultural spaces of politics could be seen as response to Butler's (2004b) call to increase the possibilities for a livable life for those who live on the margins by guaranteeing the marginalized access to the very spatiality of politics as political intelligible subjects. Through the performative expansion of the norms that constituted the spaces of politics of an agonistic and intercultural radical democracy, the door of the law would be literally opened to the man (and woman) from the country in Kafka's parable.

TOWARDS A POLITICAL GEOGRAPHY OF CHANGE

In order to make the notion of performative space productive for political geography, I have argued that we need to engage further with the performative articulations of power (relations) constituting the spaces of politics. I have suggested that closer attention needs to be paid to the conceptualization of hegemonic and counter-hegemonic processes that define or contest a spatial political order through their citational sedimented practices, in order to embed the notion of performative space within the context of politics. Framing power in antagonistic terms, the concept of performative space gains potential for a critical political geography in three main aspects: First, recognizing that the 'political is linked to the acts of hegemonic institutions' (Mouffe 2005b: 17), the (historical) citational practices – like contract theory, colonial tributes, laws and architecture – that have established a hegemonic political order can be de-naturalized and de-colonized with the Butlerian tool of performativity. Second, employing the concepts of subversion (Butler) and disarticulation (Mouffe), it can be shown how hegemonic orders are contested through slippages in the reiteration of citational practices – on the level of the subject – and through the articulation of collective demands and alternative imaginaries. Bringing these two theoretical tools together bridges collective and individual forms of resistance, and reframes the relationship between social structures and agency. The empirical analysis of counter-hegemonic struggles of indigenous movements in Ecuador highlights how change is materialized in more inclusionary spaces of politics – through slippages in the citational practices like a *kichwa* speech in a post-colonial, *mestizo* space of politics and through the collectivization of demands like the call for land reforms and suffrage. Third, Mouffe and Butler both search for alternative, more inclusionary modes of social and political conviviality. I propose that the political utopia of *interculturalidad* as incorporated in the Ecuadorian constitution can be considered as a political imaginary that both responds to Mouffe's call for an agonistic democracy and Butler's demand to expand the norms of (political) intelligibility. A critical political geography therefore should balance the need to look back and ahead. Back, at the genealogies of spaces of politics through a performative lens in order to de-naturalize and de-colonize the hegemonic 'nature' of exclusionary spaces of politics. Ahead, in order to think political horizons such as *interculturalidad* as agonistic imaginaries of a pluralist democracy.

4.
A VISUAL ETHNOGRAPHY OF POLITICAL PERFORMANCES

'Pictures are a way that we structure the world around us. They are not a picture of it' (Worth 1981, 182).

OPENING CREDITS

Figure 11: Electoral march *Figure 12: Emma waving* *Figure 13: Constituents*

See video 1: Pachakutik's electoral march, http://youtu.be/KNflPx3ZsHs
Source of video and pictures: Author

I am standing on the sidewalk of the central boulevard of the Amazonian town Francisco de Orellana, called by most inhabitants only 'Coca', and film 'my women' in the electoral march they have organized to close the electoral campaign of the indigenous and *campesino* party *Pachakutik*. Mayor Anita Rivas heads the march, energetically shouting the electoral chants (Figure 11); Emma, a candidate for the town council, waves to me when she spots me (Figure 12) and many women follow with flags and banners to support 'their' candidates (Figure 13). When I say 'my women' I don't want to sound possessive, but wish to express the emotional bond I have developed with the female politicians and candidates of the party *Pachakutik* in Coca over the time I spent interviewing and accompanying them in their everyday political activities. Hence, while filming the march on behalf of Anita Rivas, I am as euphoric as they are about the mass of people who have joined the march and the joyful atmosphere that the rainbow-colored *Pachakutik* flags and the chants produce.

What does this short visual ethnographic narrative tell about the geographies of local politics in Ecuador? What first catches our visual attention is the rainbow color of the indigenous *whipala* flag. The omnipresence of the flag's rainbow colors, so central for the constitution of this temporary political spatiality of the campaign, can be seen as a visual marker of the indigenous movement's political empowerment process. Hence, the *whipala* flag can be seen as a symbol of the indigenous movement's long struggles to acquire political voice and access to spaces of politics from which indigenous people were de facto excluded until the 1980s. What draws attention at a second glance is that the march is headed by a woman, Mayor Anita Rivas who is running for re-election and is surrounded by the other candidates of *Pachakutik*. The strong presence of women in this local political event is directly linked to the long struggle of women's organizations, which culminated in a gender-quota law in the Constitution of 1998. Hence, the visual narrative reflects recent political transformations regarding the way the local spaces of politics are gendered and ethnicized.

This introductory visual narrative on the one hand aims to present the central issues of my research about current political-transformation processes with regard to women's and indigenous people's electoral participation. On the other hand, it serves to show that visual methodologies constitute a fruitful tool 'to capture and foreground the processes of [...] change' (Spencer 2011: 34). This visual narrative highlights that spaces of politics are not only brought into being through language but also through a number of non-textual practices such as walking, smiling, clapping and symbols such as the *Pachakutik* flag. Hence, when researching processes of change and their spatial materialization, the focus needs to be on both on the linguistic performances and on the non-textual practices that constitute the very processes of change. 'Performative geographies' is the term for the research perspective that brings linguistic and embodied performances together. I argue here that visual ethnography is an especially suitable methodology for capturing these performances that are central to performative geographies.

GENRE: VISUALIZING THE PERFORMATIVE TURN

At the beginning of the millennium, Nash (2000: 654) identified a 'metaphorical and substantive turn from "text" and representations, to performance and practices' in human geography. The 'performative geographies' that result from this turn criticize poststructuralist geographies for remaining 'strongly focused on the verbal' (Crang 2002: 652) and call for a stronger focus on the embodiment of everyday practices and the cardinal role of the body within these practices. The concept of performativity, as developed by Judith Butler (1990a), offers a possibility to combine the performances and practices – the 'acting out' – with the enabling and disciplining effects of discourse.

Butler has shown that one's gender identity is neither solely located in bodily difference nor is it the result of a free-floating choice. Rather, women and men learn to perform the sedimented forms of gendered social practices that become so routinized as to appear natural. Gender does not exist outside its 'doing', its performance, but this performance is also a reiteration of previous 'doings' that become naturalized as gender norms. In an attempt to integrate Butler's thinking into human geography, Gregson and Rose (2000: 434) have argued that 'space too needs to be thought of as brought into being through performances and as a performative articulation of power'.

There are vibrant discussions in human geography as to how to methodologically redeem the theoretical ambition of performative geographies (Davies and Dwyer 2007, Dirksmeier and Helbrecht 2008, Latham 2003, McCormack 2005, Morton 2005). Visual methodologies are considered a fruitful way out of the 'textual trap' (Lorimer 2010: 242) as they permit a focus on embodied practices which escape text- and talk-based approaches. Both anthropology and human geography have a long tradition of using photographs, diagrams, maps and film. Rose (2011: 10), however, points out that a proliferation of visual methodologies has occurred only in recent times within and beyond these disciplines (El Guindi 2004, Emmison and Smith 2000, Garrett 2011, Pink 2001, Rose 2011, Spencer 2011, Sturken and Cartwright 2009). Human geographers have increasingly incorporated visual methods such as video (Garrett 2011, Kindon 2003, Simpson 2011) or photography (Latham 2003, Latham and McCormack 2007, Rose 1997a, 2004) in their research in an attempt to pay greater attention 'to the corporeal aspects of knowledge production by focusing on [...] body language, use of and movements through space' (Kindon 2003: 147). For Garrett (2011: 12), 'there is a great deal that cannot be written or spoken that can be expressed through performance, gesture and polysemous representation on film'. Hence, the strength of visual methods consists of capturing non-linguistic performances that have gained importance in the theoretical concerns of performative geographies.

While the increasing use of visual methods in human geography can be seen as one methodological response to the performative turn, Herbert's (2000) 'call for ethnography' can be considered another attempt to enrich the methodological tool kit of human geography with methodologies that are suitable to 'explore the [non-linguistic] tissue of everyday life' (Herbert 2000: 551). His request 'for [more] ethnography' in human geography has gained increasing support (Kofman 2008, Kuus 2012, Megoran 2006, Müller 2012, Schurr and Kaspar 2013, Woon 2013). Especially as a consequence of the turn from texts and representations to performances and practices (Nash 2000: 654), geographers have recognized that ethnography 'provides singular insights into the processes and meanings that undergird socio-spatial life' (Herbert 2000: 550). Megoran (2006) highlights that these particular insights are possible due to the epistemological distinction between ethnographic and social-science research. For Verne (2012) this epistemological distinction is grounded in Geertz's (1973: 6) understanding of ethnography as an 'intellectual effort' and not the mere use of certain techniques. The 'intellectual effort' of doing ethnography consists of 'trying to read (in the sense of 'construct a reading of') a manuscript – foreign, faded, full of ellipses, incoherencies, suspicious emendations, and tendentious commentaries' (Geertz 1973: 10). In order to be able to 'construct a reading of' this foreign, unfamiliar manuscript, a long-term engagement and close observation of the researched context is necessary. It is important to highlight here that ethnography as a research methodology is inherently visual as it focuses on practices of looking and observing. While the 'visual in ethnographic research has generally not been used intrinsically for interpreting and representing ethnographic data' (O'Neill et al. 2002: 72) but as merely illustrative, visual ethnography makes the visual aspect of ethnography explicit by giving the visual data a more analytic treatment.

Although visual and ethnographic methodologies have received increased attention in human geography, empirical work that brings both methodological approaches together in the form of visual ethnography is still scarce. Byron's (1993) study about post-war migration in the Eastern Caribbean is an early example of geographers integrating video in their ethnographic research. More recently, Kindon (2003: 142), who, during her ethnographic fieldwork, conducted a participatory video project with members of a Maaori tribe in Aotearoa, New Zealand, advocates the use of participatory video making as a feminist methodology that provides a 'practice of look "alongside" rather than "at" research subjects'. In a similar vein, Young and Barrett (2001) highlight the effectiveness of using visual 'action' methods to encourage child-led activities in their research with street children in Kampala, Uganda.

I argue here that human geography can benefit beyond this participative aspect from anthropology's experiences with visual ethnographies (Collier and Collier 1986, Mead 2003, Pink 2001) when linking visual ethnographic methods to current theoretical discussions about performativity and performance in human geography (Gregson and Rose 2000). Visual ethnography offers a unique possibility to connect observed performances to performativity – the discursive framing that disciplines and enables these performances. To bring together the study of performance and performativity, a visual ethnography needs to be developed that combines visual research methods with insights gained from studies of visual culture as suggested by Rose (2011). In the following, I develop what I call a 'visual ethnography of performativity' that explicitly connects observed performances and performativity – the 'citational practices' (Butler 1993: 108) which enable and discipline subjects and their performances. Hence, such a visual ethnography of performativity focuses on the relation between subjective performances and hegemonic discourses in which these performances are embedded in. Pink (2008) suggests a 'media ethnography', the close examination of visual representations in media, as a way to engage with hegemonic visual representations within certain cultural contexts. By linking self-produced visual data with visual data produced by the media, it is possible for the researcher to relate, compare and contrast his or her own visual data with hegemonic visual representations of certain performances.

By drawing on my visual ethnography of politicians' identity performances in Ecuador, I will show in the empirical section of this chapter how the filmed identity performances can be linked and contrasted to hegemonic discourses around masculinity, femininity, whiteness, and indigenousness represented in Ecuador's visual culture. Before doing so, I would like to reflect on the way the video camera became not only a research device but an active agent that enacted reality within my research process in a particular way (Law and Urry 2004).

STAGE DIRECTION: STRUGGLING WITH THE CAMERA IN THE FIELD

In this section, I outline my methodological approach and reflect critically on the research process and related ethical concerns. It is worth highlighting that originally the research project did not include visual methods. Identity and spatial constructions within local politics were 'intended' to be captured during extensive ethnographic fieldwork in three municipalities mainly through interviews and (participant) observation. At the beginning, I was reluctant to use video or photography. Postcolonial critiques (Said 1978) about the way the East became orientalized by the West through images weighed heavily on my post-colonial academic consciousness. My firm ethical conviction was rocked when during my first field trip I accompanied a mayor and a television cameraman to a political meeting. Watching the mayor preparing herself for her performance by putting on make-up and tidying up her clothes, I felt that a camera might even trigger what I was looking for: The (un-)conscious performance of gendered, ethnic, classed and political identity. I returned for my second field trip to the same women with a digital camera and video camcorder and asked timidly for permission to photograph and videotape their daily activities. I was overwhelmed about how euphoric every single woman was about the idea and all of them asked me to allow them to use the recordings later for their own purpose. From this moment on, I was given a new role: I became the personal camerawoman and was constantly asked to accompany them to meetings, audiences and gatherings. At last – in my view and probably also in the view of my research partners – I had a reason to be around the politicians all the time. The up-coming electoral campaign literally turned out to be the ideal stage to research the politicians' performances. Alongside local media people, I recorded political speeches, electoral demonstrations and house-to-house campaigns.

During my research process, the camera functioned as a kind of 'can-opener' as described by Collier and Collier (1986): First, playing the camerawoman put me in an ideal position to observe the women I was researching and gave me legitimacy to do so. Second, when I showed the videos and photographs to the politicians, fruitful discussions took place about the course of the event, their feelings during the event and reflections about their performances. Even though I had lived in Ecuador for a long time and had worked with local politicians on various occasions, my own position as a middle-class European researcher often 'directed my gaze' in an Eurocentric way to certain aspects (often the 'exotic') that did not attract the attention of the research participants (e.g. indigenous symbols such as the *Shigra* bag) and that at the same time overlooked other aspects (e.g. small gestures between the candidates that they directly interpreted as signs of friendship or conflict). The constant discussion about the videos and photos with research participants and later on with Ecuadorian colleagues was crucial for me. I could thus challenge my biased interpretation of the visual data and avoid the risk of producing interpretations that exoticized or stereotyped those I was researching (Spencer 2011: 489).

In hindsight, I think it is important to acknowledge both the added value and the challenges of using a camera in ethnographic research. Pink (2001: 98) reminds us that the 'reflexive use of video in ethnography means using video not simply to record data, but as a medium through which ethnographic knowledge is created'. Within my research process, the camera was used to collect data to then illustrate my text-based arguments, but more importantly, it played an active role in producing the data – both in its visual forms, when the camera provoked certain performances from research participants, but also in textual forms, through the discussions when I watched the visual data with research participants. Further, the 'media ethnography' (Pink 2008) I conducted by analyzing local and national newspaper articles, television programs and electoral propaganda facilitated embedding my own filming and the analysis of my visual data in the visual culture of Ecuadorian politics.

Nevertheless, during the whole research process, the awareness of visual anthropology's colonial roots frequently made videotaping a contested endeavor for me when confronted with ethical questions about whom and when to film, when feeling power relations between myself and those filmed due to the material value and social status of my camera and finally when wondering whether my own position complicated and biased the filming (Kindon 2003). On many occasions, I decided not to film and preferred to silently observe what was going on. I often decided spontaneously and intuitively when and where to film. I would also often check with some women politicians I had already established rapport with as to whether it was appropriate to film. In general, it was less problematic to film the politicians themselves, as they are used to being at the center of media interest, than to film farmers or indigenous people who are rarely faced with a camera.

ON THE SET: IDENTITY PERFORMANCES ON THE POLITICAL STAGE

What do we see, then, when we look through the camera at women politicians' identity performances from a performative perspective? Looking at the case of Guadalupe Llori, Prefect of the Amazon Province Orellana, I demonstrate how the concept of identity can be examined through the lens of visual ethnography. I also analyze her performances during the final campaign day in order to discuss the way her performance is embedded in and at the same time challenges hegemonic imaginaries of the political that are (post-)colonially gendered and racialized as masculine and white. Although much more could be said, I will restrict my analysis to the three central analytical categories in my research project: *political identity, ethnicity*, and *gender*.

Figure 14: Guadalupe Llori on the stage. Source: Author

During the observed campaign event, Guadalupe Llori is staged in the middle of the scene, which emphasizes her central political position within the indigenous party *Pachakutik* (Figure 14). She is wearing light casual trousers, sandals, a yellow farmer's shirt and a baseball cap. She is outfitted with accessories in the rainbow colors of *Pachakutik,* which resemble the Incan *whipala* flag. An indigenous handbag, the *shigra*, is slung over her shoulder.

What does the way Guadalupe is dressed tell about her *political identity*? While all candidates wear small accessories in rainbow colors, Guadalupe is covered from head to toe with these accessories, including earrings, bracelet, scarves and a ribbon around her baseball cap. This attire she has chosen allows the people to understand that *she* is the real and only representative of the *Pachakutik* movement. By wearing quite casual clothes, including the *shigra* handbag, she positions herself close to the people, *her* people and *her* community. While traditionally white and *mestizo* male politicians are dressed rather formally with a suit, more and more politicians of the left have replaced their formal clothes by a casual look.

Not only through her dress, but also through her political rhetoric, Guadalupe performs her political identity, when, for example, in the first few minutes of her speech she emphasizes (Video 2: campaign speech Guadalupe Llori, http://youtu.be/KNflPx3ZsHs) that

> 'we must not divide the people (pueblos), we must not divide the indigenous nationalities (*nacionalidades*), we must stay together and fight for balance (*equilibrio*), equality (*equidad*), *compañeros*' (political speech 21.04.2009, own translation).

Without elaborating her ideas as to how this balance and equity should be achieved, she picks up a rhetoric found in the national indigenous organization

(CONAIE) and a worldview celebrated by indigenous leaders. She states in an interview:

> 'Because I have identified since childhood with my indigenous brothers and have incorporated their ideological principles and moral values, I chose to join *Pachakutik* in order to fight for the impoverished mass of my people' (interview Llori, 17.02.2010, own translation).

While the performance observed so far might suggest that she is performing an indigenous *ethnic identity*, in fact she does not have biographical roots in an indigenous community. She compensates her inability to speak in *kichwa,* the indigenous language, by decorating herself all over with indigenous symbols and by emphasizing in her speech her political proximity with indigenous values. Within the Province of Orellana, people picture her as a typical *mestiza* who supports the cause of the indigenous people. Guadalupe Llori herself, however, appropriates different ethnic performances according to the audience and the message she wants to deliver. The video from the Human Rights Foundation Forum where she was honored for her fight against petroleum activities in the province shows that in the video she refers to herself as indigenous (Video 3: Guadalupe Llori at the Oslo Freedom Forum, http://youtu.be/KNflPx3ZsHs).

Figure 15: Guadalupe Llori's campaign poster
Source: Author

Her campaign poster (Figure 15), however, represents her in a hybrid mixture of a white upper-class woman – emphasized by her dyed blonde hair and her make-up – and an indigenous identity expressed through the rainbow colored earrings and scarf. This image might be interpreted as an attempt to respond to discursive hegemonic imaginations of a politician as white, well-educated, and upper-class and to establish a close relationship to the 'simple' poor and indigenous people at the same time. Hence, Guadalupe's performance reveals an ambivalent and hybrid ethnicized political identity.

What can be said about Guadalupe Llori's performance of *gender identity* when contrasting the visual recordings of her electoral speech with this election poster? While the election poster emphasizes her femininity through the earrings, make-up, lipstick, and her well-coiffed hair, her embodied performance on the political stage, including her clothing, body movements and the tone of her voice enacts a rather masculine performance. Listening to her campaign and watching her body language reminded me of other socialist (male) political leaders such as Hugo Chavez, Fidel Castro or Che Guevara. Her affinity to these men becomes evident by wearing a wristlet with the icon of Che Guevara or by using a similar rhetoric such as *'hasta la victoria siempre'*. By doing so, she embeds her own identity performances within socialist imaginaries of politics that dominate contemporary Latin American (left-wing) politics (Castañeda 2006). Her body language and way of speaking copy the performances of her male idols when she shouts energetically or raises her left index finger (Figure 16) like Che Guevara (Figure 17) or Fidel Castro (Figure 18).

Figure 16: Guadalupe Llori *Figure 17: Che Guevara* *Figure 18: Fidel Castro*
Source: Author *www.randomfactsoftheday.com* *www.vorwaerts.ch*

This description of Guadalupe Llori's performance on the political stage has focused so far on the question of how Guadalupe is 'doing identity' (West and Zimmerman 1987) through her body movement, clothes, use of symbols, and rhetoric. The performative dimension of her performances now needs to be considered by questioning the subversive and transformative power of new political subjects like Guadalupe Llori, a power that parties like *Pachakutik* claim for themselves. Because 'the notion that identity is performative (and note that this does not imply voluntarism) lends itself to claims about doing gender differently [...] this provides opportunities for radically redoing gender' (Cream 1995: 39). What Cream claims along with Butler (1990a) about gender identity is certainly also relevant for other identity constructions like political identity or ethnic identity. With regard to transformations of the political towards a more inclusive and participative mode, it needs to be asked to what extent new political subjects 'redo' political subjectivity and concomitantly the imaginations about and constructions of the political. As the empirical examples suggest, there is no easy answer to this question. Guadalupe Llori's ambivalent identity performances are typical for many female and indigenous politicians. In their performances, new political subjects constantly negotiate between traditional imaginations about how to do politics and be a politician and, at the same time, distance themselves from and subvert these same discursive imaginations. Whereas Guadalupe Llori challenges the masculinity of Ecuador's postcolonial political spaces through her presence, her jewelry, and her rhetoric on the importance of her family and her love for her people, she reproduces masculine stereotypes about politicians by imitating masculine body movements, gestures and rhetoric. The same can be said with regard to the discursive imaginations of politics as 'white people's business'. Guadalupe challenges these imaginations through the use of indigenous symbols and rhetoric but she still clearly reproduces the whiteness of the political sphere by emphasizing her own whiteness in her campaign poster.

CLOSING CREDITS

Visual ethnographies have been advocated here as a productive methodology to redeem the theoretical ambition of performative geographies. I have argued along with Rose (2011) that in order to conform to the theoretical assumption that performances are always linked to performativity (Gregson and Rose 2000), the collected visual data needs to be analyzed against the backdrop of the visual representation of hegemonic imaginaries. Such a visual ethnography of performativity combines the strength of visual and ethnographic methodologies to capture both performances and the performative discourses framing them. Performative geographies can benefit from insights gained in sociology and visual anthropology, both of which have a long tradition of focusing on the body as a rich source of visual data. The suggested visual ethnography of performativity integrates these insights into human geography and demonstrates that visual methodologies are a productive way to capture and analyze non-linguistic, embodied performances. Visual ethnographies in this sense are an appropriate methodology for geographies of embodiment (Harrison 2000), affect/emotion (Pile 2009a) and everyday practices (Simonsen 2007) that are interested in the way spatialities are brought into being through non-linguistic practices such as feelings, body movements, non-verbal interactions, clothing, and symbols.

Performative geographers aim to understand the way the observed performances are embedded in performativity, the discursive framing that enables and disciplines performances. To meet this aim, a visual ethnography of performativity needs to reveal a profound understanding of the way hegemonic discourses around, for example, femininity or whiteness are visually represented. To identify and analyze hegemonic discursive imaginaries, ethnographic methodologies offer great potential because they focus on an emic understanding of social life and long-term embeddedness in a research context. Hence, visual ethnographies that take into account the discursive framing of observed data appear to be a promising methodology for geographic research seeking to bring together practices and discourses (Müller 2008) in order to develop more nuanced and vivid accounts of social life.

The empirical example was aimed to show how the combination of different visual, ethnographic and text-based methods applied in this research project offered a way to capture the ambivalence in the new political subjects' identity performances and situate them within hegemonic discourses that define the political as masculine and white. Transformation processes such as the one described in the case of Ecuador's local spaces of politics frequently consist of slight slippages and variations of the hegemonic scripts. To identify and interpret these slight embodied slippages, such as the use of indigenous clothing or feminine or indigenous ways of performing political subjectivity through body language, a visual ethnography of performativity as developed in this chapter is especially well suited. The particular strength of visual ethnography lies in capturing 'seemingly unremarkable signs of everyday life' (Spencer 2011: 47) that are easily overlooked when one is taking ethnographic notes and absorbed by the event itself. Visual note taking offers the possibility of examining past activities as 'formerly present', immersing oneself again in the event but focusing on small hints and deviations from hegemonic ways of performing political subjectivity and spatiality.

To sum up, I would like to highlight that a visual ethnography of performativity as developed in this chapter not only serves to capture the embodied performances that performative geographies are concerned with, but also highlights the performative aspect of any research process (Gregson and Rose 2000, Pratt 2000a). It does so by making the gaze of the researcher more explicit (Kindon 2003: 145) than in conventional ethnographic research. Visual ethnography, by allowing the researcher and participants to discuss visual data together, also offers the opportunity to 'see together [with the research participants] without claiming to be another' (Haraway 1991: 193) and produce in a collaborative research process a 'negotiated version of reality' (Pink 2001: 24).

EMOTIONS

5.
PERFORMATIVE EMOTIONS IN ELECTORAL CAMPAIGNS

'The world of politics is inevitably and rightly a world full of emotions' (Clarke, Hoggett, and Thompson 2006: 5).

QUE VIVA PACHAKUTIK: EMOTIONS IN LOCAL CAMPAIGNS

It is a very hot day here in Coca, the Capital of the Amazonian Province Orellana. I have accepted an invitation from Mayor Anita Rivas to participate in the closing campaign of the indigenous and *campesino* party *Pachakutik*. In the midst of a euphorically cheering crowd, I stand at the foot of a temporarily erected stage and listen to the electoral speech of provincial Prefect Guadalupe Llori: 'We must not divide the indigenous peoples, we must not divide the [indigenous] nationalities, we have to stick together to maintain equilibrium [with nature], to maintain equity, *compañeros,* because success is here already. To love and to serve *el pueblo* until it hurts and the more it hurts, the more I love you *compañeros*. This is *Pachakutik*, list 18! [...] We waste our time, quarrelling over a president who only takes notice [of our province] when we strike and when he needs oil, *compañeros*. That's why I invite all of you to consider carefully and decide which local authorities you trust, who [unlike other candidates] don't take advantage of the fact that a president, a presidential candidate is popular, trying to get onto the [party] lists like *Waira pamushkas*[13], trying to fish [for votes] to get elected, sometimes they are here only for a year or two, they think they can save our *pueblo*, *compañeros*, they even come to criticize our history of suffering, of sorrow, sometimes even of distress, *compañeros*. This isn't right, that's why you have to think about who you vote for!' Spontaneously I lift up my hands with the others, chanting with the women and men around me *'Lupita, Lupita', 'Que viva Pachakutik', 'Que vivan las mujeres'* [14] (ethnographic note, 21.04.2009,Video 4: 'Guadalupe Llori', http://youtu.be/4ShQZcjhmsE).

---

13 In 2008, indigenous Mayor Auki Tituaña called President Rafael Correa 'the Waira pamushka of Ecuadorian politics'. He regards Rafael Correa in the same way as he views the Spanish conquistadores: as 'Waira pamushkas' (literally: son of the wind, metaphorically: people who appear out of nowhere) who 'offers change, development and well-being, but in reality he just brings neoliberalism, partydocracy (sic!) and the same things as always' (Ecuadorenvivo: 22.09.2008).
14 In an attempt to problematize the political act to speak for/with others (e.g. Alcoff 1991), I decided to make my agency as translating geographer visible by using a 'translation which keeps words in the source language as a visual marker of indeterminacy' (Müller 2007: 212).

Figure 19: Guadalupe Llori's campaign speech *Figure 20: Crowd cheering*
Source: Author *Source: Author*

The reconstruction of this local campaign event raises interesting issues about the role of emotions in local politics. First, it emphasizes that emotions are a key ingredient of political campaigns. Political scientists, psychologists and sociologists have provided empirical evidence that emotions displayed in electoral campaigns influence and motivate citizens to participate and vote (Marcus 2002, Marcus, Neumann, and MacKuen 2000, Neumann et al. 2007). The growing body of literature about emotions and electoral politics is dominated by experimental (Ansolabehere and Iyengar 1995), quantitative (Clarke et al. 2009) or neuroscientific approaches (Marcus et al. 2006, Marcus, Neumann, and MacKuen 2000) that show the impact of campaigning on the constituents and their electoral choice. While these studies underscore that 'emotions in campaigns do matter' (Clarke et al. 2009, Hillygus 2010, Johnston and Pattie 2006), vivid accounts of the emotions in electoral campaigns and the way they are related to certain identities remain a lacuna within this body of literature. In order to understand the way the political is produced as an emotional space, this chapter addresses this lacuna by focusing on the way emotions are (re-)produced by political candidates rather than focusing on the emotions of constituents. By doing so, this study focuses on the 'doing' of emotions that is frequently taken for granted in political science literature, which largely concentrates on the impact of emotions on the constituents. Employing an ethnographic approach that closely observes the emotions on display on the electoral stage, the chapter contributes to this body of literature by giving a detailed account of the emotional performances that create the very spaces of campaigning. Drawing on Gregson and Rose's (2000) notion of performative space, I argue that the spaces of campaigning are brought into being through emotional performances. The performances of the candidates, their electoral speeches, their interaction with supporters, as well as the heterogeneous relations between their performances and non-human agents such as microphone, political banners or the stage itself, turn everyday spaces – whether a plaza, school or street – into spaces of campaigning.

Second, the narrative shows that, within the spaces of campaigning, emotions are closely linked to certain identities such as indigeneity ('we the indigenous peoples') or local belonging ('our *pueblo*'). While studies in political science have addressed the effects of electoral candidates' gender or ethnic identity on voter turnout and election results (Herrnson, Lay, and Stokes-Brown 2003, Marschall 2010, Sanbonmatsu 2010, Stokes-Brown and Neal 2008, Warf 2011), little is said about the role of emotions in the processes of gendering and racialization of electoral spaces. The electoral speech of Guadalupe Llori, cited above, provides an example of how these processes of gendering and racialization take place. In her speech she recalls emotions associated with memories of shared experiences, such as the political neglect of the province or the discrimination of indigenous people. Paying attention to the ways emotional performances are linked to these collective experiences, I respond to Tolia-Kelly's (2006: 215) call for an emotional geography that acknowledges 'power geometries of our present as linked to our past' that 'effect the parameters and flows of affectual capacities and sensitivities'.

Third, the political opposition Guadalupe Llori addresses in her speech is made up of white and *mestizo* masculine elites personified in left wing, *mestizo* President Rafael Correa. By doing so, a gendered and racialized binary is constructed which lies at the basis of a distinction between 'good' and 'bad' ways of doing politics. This binary draws on the promises of female and indigenous politicians to do politics differently (Cañete 2004b, Van Cott 2008) from white and *mestizo* men who have dominated Ecuadorian politics during the last century (Becker 2003). Women, indigenous and Afro-Ecuadorian people are considered to be new political subjects as they have only gained significant political visibility since Ecuador's return to democracy in 1979. The term new political subjects results from their participation in new social movements (Alvarez, Dagnino, and Escobar 1998) through which they have fought for their political rights. Due to the post-colonial situation of Ecuadorian politics resulting in the recent integration of formerly excluded social groups, Ecuador presents an interesting case by which to evaluate the emotional impact of new political subjects on the production of political spaces. In Ecuadorian politics, constituents assess the political performances of new political subjects against the backdrop of hegemonic political subjects, which are predominantly *mestizo* populists. Populism in its different guises of classic populism (1940s-1970s), neoliberal populism (1990s) and radical left-wing populism (since 2007) has dominated the Ecuadorian political stage during the last century (de la Torre 2010). By so doing, it embodies the hegemonic political style that serves as the background against which the novelty of new political subjects is judged. Populism has also defined the emotional atmosphere of Ecuadorian politics through its particular political style. In fact, 'populist rhetoric radicalizes the emotional element common to all political discourses' (de la Torre 2010: 4), by constructing politics as an emotion-laden, antagonistic struggle between *el pueblo* and *la oligarquía*.

THE PERFORMATIVITY OF EMOTIONS IN POLITICAL SPEECH

From a conceptual perspective, my analysis of the emotional geographies of campaigning is especially interested in questions of continuity and change with regard to Ecuador's recent political transformations. In order to be able to conceptualize these dynamics of continuity and change within electoral politics, I propose a performative understanding of emotion. By so doing, I shift the focus from what emotion is – a question that dominates the current debate around emotional/affectual geography (Pile 2009b, Thien 2005) – to what emotions do to alter or reproduce social relations. Combining Sara Ahmed's (2004a, 2004b: 82–100, for a good summary see Scharff 2011: 217–218) concept of performativity of emotion with insights from Samuel Hampton (2009) about the performativity of political speech, I develop below a conceptual framework that enables analysis of emotions as outcomes of speech and body acts performed on the political stage.

Ahmed reflects on the performativity of emotions by looking at how certain emotions such as disgust, shame or love involve not just body acts, but also speech acts. She understands performativity according to Judith Butler (1993) as the power of discourse to produce effects through reiteration. The politics of performativity require the iterative power of discourse to produce a certain emotion, since 'emotion does not exist before one says something, i.e., before producing a speech act' (Belli, Harré, and Íñiguez 2010: 259). It is important to acknowledge that for Butler, a complete performance requires a combination of both speech and body acts. As Belli et al. (2010: 260) point out, '"I love you" is an expression of the body as totality, not just a simple phrase'.

The emotions generated in the contact between different bodies depend then on histories of associations. Drawing on the colonial encounter between white and black bodies, Ahmed (2004b: 92) shows that signs such as disgust or hate become 'sticky' through repetition and 'some objects become stickier than others given past histories of contact'. These past associations that 'stick', to use Ahmed's coinage, certain emotions to certain bodies – such as the hate against the postcolonial master – make emotional performances performative: 'in reading the other as being hateful, the subject is filled up with hate, as a sign of the truth of the reading' (Ahmed 2004a: 32).

Ahmed's idea that emotions are performative because they repeat past associations and by doing so generate their object is crucial for my analysis. The new political subjects' emotional performances frequently recall certain emotions (such as people's disappointment about the neglect of the province by political authorities) through aligning these past (post-)colonial associations with white/ *mestizo* and masculine bodies. Referring to President Rafael Correa as *Waira pamushkas*, for example, can be considered as a speech act that seeks to evoke emotions by associating him with past colonizers. In summary, Ahmed's performative notion of emotion shows how emotions align with bodies (the racist colonizer); how these emotions come to 'stick' to certain bodies (the fear of indigenous people of racist discrimination); and how these associations move subjects to behave in specific ways (e.g. indigenous people avoiding contact with racist *mestizo*s).

Having briefly summarized the main points of Ahmed's notion of emotions as performative, I now develop insights from Hampton about the performativity of political speech. Hampton (2009: 4) argues that 'as linguistic, embodied, immanent and affective performances, political speeches are the paradigm of performative politics'. Building on Butler's theory of performativity, Hampton considers performance as a central feature of politics, as the subjectivity of the politician and the spatiality of politics do not pre-exist the performance. Rather, the very subjectivity and spatiality of politics are produced through the performative act itself. For Hampton, scripts and storylines are the devices of political performativity. They provide 'the materials and substance for the ephemeral processes of circulating reference that characterizes emergent political performances' (Hampton 2009: 16). Storylines serve to reduce complex political issues to a simple plot in a (geo-)political[15] drama. Scripts are compiled of hegemonic routines, narratives and vocabularies employed by political subjects in their performances. Through processes of citation and reiteration, certain storylines and scripts become embedded and transformed into hegemonic political norms and imaginaries. These hegemonic norms are constantly reproduced in citations that circulate through the past, present and future. To look at political speeches from a performative perspective offers the possibility of examining how speech and body acts reproduce and/or challenge these hegemonic emotion-laden scripts and storylines. Similar to how Ahmed shows how certain emotions 'stick' to certain bodies through past associations, emotions also 'stick' to dominant scripts and storylines.

Speeches held during political campaigns provide a fruitful basis for examining the emotions that constitute political spatialities for two reasons: First, political speeches are a historical constant of electoral politics and can be considered as one of the political acts that most explicitly seek to evoke emotions. Second, political speech can be understood as a discursive device which employs performative techniques such as storylines and scripts with the aim of recalling past and present emotions that are linked to memories of political success and/or suffering. In so doing, political speeches produce, alter, and consolidate these memories. With their constant citation of styles, techniques and certain phrases used in past political speeches that are intended to evoke certain emotions in the audience, political speeches are an ideal site to study both the continuities and transformations of politics. Taylor (2003: 6) further reminds us that the political 'performative is not simply an act; it is a pact'. Hence, the political speech act that successfully evokes emotions in the audience 'requires the coalescence of an audience, without which the utterance has no resonance and thus no performative power' (Hampton 2009: 7). In the same way as Austin (1962: 14–15) shows that speech acts are only felicitous if a conventional procedure (such as a campaign script) exists that guides the action, performative emotions rely on previous norms and conventions of speech

---

15   I use the parentheses to indicate that I refer here both to geopolitical and political storylines. I define geopolitical storylines as experiences that emerge from international political relations and violence whereas political storylines refer to political processes taking place within national boundaries.

and body acts to generate the object that they name (for example the menacing (post-) colonial white elite). For a speech or body act to become performative, it needs to be spoken or enacted to others, 'whose shared witnessing [of the menacing white men] is required for the affect to have an effect' (Ahmed 2004b: 94). An audience is required for a speech or body act to be felicitous, as the audience needs to repeat the attribution of a certain feeling to an object articulated in the speech act. The speech act not only generates the object (the menacing white men) but also a community of those who are bound together through shared feelings (e.g. of fear).

RESEARCHING EMOTIONS

Understanding emotions as performative requires focusing on the ways in which both linguistic and embodied performances produce an affect that has an effect on others (Ahmed 2004b: 94). How can such a performative approach to emotions be methodologically justified? There is currently wide debate in the field of emotional/affectual geographies on how to capture emotions and affect in empirical research (Davies and Dwyer 2007, Morton 2005, Rose 2004, Simpson 2011, Wood and Smith 2004). Three main approaches can be identified with regard to the way geographers engage with the emotional dimension of social life: interviewing, ethnography and visual methods. In the following, I present each of them and show how I have employed them within my research project.

First, qualitative methods such as interviewing or (participant) observation are helpful to capture the way people talk about their emotional experiences. From a performative perspective, this kind of audio data serves to analyze how interviewees' linguistic emotional performances draw on, and are constrained by, social norms and hegemonic imaginaries, such as in my own research the emotionality of populist campaigns. At the same time, these norms and imaginaries define which emotions are expressed by whom and how they are reproduced through their own emotional performances (Wiles, Rosenberg, and Kearns 2005). In analyzing these linguistic performances, it is crucial to pay attention not only to *what* people say, but also *how* they talk about emotional experiences. To this end, forty interviews with local female politicians[16] and around twenty political speeches

---

16 Semi-structured and partly biographical interviews were conducted with forty local female politicians, among them seven provincial candidates for the national assembly, three (vice)prefects, three (vice)mayors, thirteen town councilors, fourteen rural councilors as well as with fifteen women heading local women's organizations. With regard to the ethnic composition of the interviewees, eight of them self-identified as Afro-Ecuadorian, twelve as indigenous and the rest considered themselves to be *mestizas*. Their age ranged from twenty to

were recorded, transcribed and subsequently analyzed with regard to their content, the use of speech acts and the tonality of speaking. Content, (ritual) linguistic expressions, and the tonality of voice were then compared, contrasted and related to political speeches of Ecuadorian populist politicians by analyzing (video) recordings of these speeches and literature that cites extracts from these speeches.

The use of ethnographic methods is regarded as a second means of looking at emotions. Extended fieldwork can provide an invaluable opportunity of establishing close relationships in the field that are then crucial for sensing and discussing the emotions found to be present in that situation, for the purposes of research. Between 2006 and 2010 I spent a total of eighteen months in the provinces of Esmeraldas, Chimborazo, and Orellana. I accompanied two prefects, two mayors and four councilors, all of them women, for several weeks in their everyday political activities and later on in their campaigning activities, spending a week in each province during the electoral campaign in 2009. Through this long-term engagement with these women, it was possible to address emotions not merely as a object of study but rather as 'a relational, connective medium in which research, researchers and research subjects are necessarily immersed' (Bondi 2005b: 433). Hence, capturing, analyzing and writing about emotions during the electoral campaign in 2009 is a reflection of how I experienced the emotional atmosphere during campaign events, how certain emotions were transmitted to my own body through the bodies around me when talking with constituents, standing and chanting in the crowd of supporters or when observing or talking to the female politicians before, during or after their performances on stage.

Third, visual methods are increasingly applied in human geography to focus more explicitly on the embodied performances through which emotions are expressed (Garrett 2011, Simpson 2011, see also chapter 4). I made video recordings of many of the campaign events I participated in. These recordings successfully captured minute details of the emotional performances and facilitated a close analysis of the event afterwards. While I agree with Simpson (2011: 343) that the video recording 'does not necessarily present or give a sense of the affective relations present in [...] encounters', due to the possibility of presenting both visual and sonic impressions such as body movements, chants, clapping etc., visual data transmit the emotionality of certain spaces in a more detailed way than written accounts are able to do. In my case, photographs and videos recorded during field-

---

sixty-eight, and their educational background varied between two years of primary school, to having received a university degree. The women I interviewed belonged to a diverse range of political parties present in Ecuador. Interviews with constituents were not systematically realized as my focus was on the emotions displayed by the candidates. The video-recordings and informal chats with the constituents during the campaign events, however, offered me insight into the constituents' opinions about and feelings towards the candidates.

work were later used to (re-)construct a close description of events, focusing on the exact speech acts of the candidates as well as on their embodied performances. Discussion of the visual data with people in the research context and with Ecuadorian colleagues was crucial for my analysis of the emotional performances of the campaign events. To transmit an impression of the emotional atmosphere of an event, some of these recordings are presented to the reader in a linked video (http://youtu.be/4ShQZcjhmsE). Further, analyzing video recordings of Ecuador's populist politicians during campaign events enabled me to compare the performances of the new political subjects with hegemonic ways of campaigning.

Analyzing the data and writing about the emotional geographies of campaigning in Ecuador, I was confronted with the challenge of how the actual feelings of the people and their performative effects can be read or deduced from the (visual) recordings. While feelings are necessarily subjective, personal and not easily shared, analyzed and interpreted, in my view the detailed and, in particular, collaborative nature of the analysis of the recordings offered the possibility to approach the emotionality produced through political speeches. Focusing on speech acts realized in these political speeches was one way of coming to terms with the performativity of emotions that brings the very spaces of campaigning into being. Further, reflecting on my own emotions felt and experienced during these events, remembering the affective atmosphere of the bodies that surrounded me and watching the video recordings with Ecuadorian colleagues was another way to approximate the ephemeral and temporary emotional geographies of campaigning. Inevitably, the narratives and video clips (re-)narrated and shown in my analysis are partial and screened through my own 'eye/I' (Kondo 1990) and are therefore just one possible interpretation of the emotionality displayed in Ecuadorian political spaces.

EMOTIONAL GEOGRAPHIES OF POPULISM

What are the hegemonic storylines and scripts that define the emotionality of political campaigning in Ecuador? As discussed in the introduction, populism has defined the emotional register in Ecuadorian politics during the past decades and can thus be considered as the political style that has become established as the hegemonic way of campaigning. In his extensive work on populism in Ecuador, de la Torre (2001, 2010) has shown that Ecuadorian politics are characterized by the discursive Manichean confrontation between *el pueblo* and the oligarchy. The populist rhetoric employs dominantly the two emotions *rabia* (rage) and *amor* (love) in a Manichean way. In the following, I discuss how populist spaces of campaigning are brought into being through emotional performances of *rabia* and *amor*.

RABIA OF THE PEOPLE AGAINST THE OLIGARCHIES

'We have to defeat the oligarchy, the partyarchy, and the wigs who want to go back to the past' (President Rafael Correa cited in de la Torre 2010: 190).

As the extract from President Correa's campaign speech shows, the populist spaces of campaigning are produced through a rhetoric of war (use of the term 'defeat'). The campaign is turned into a 'battlefield' between the oligarchy and *el pueblo*. Through his speech act that reproaches the oligarchy for wanting to return to the past, Correa equates his opponents with the colonial oligarchy. Emotions of *rabia* are evoked in the constituents by recalling past associations with the colonial struggle between *el pueblo* and the colonial elites.

It was President Velasco Ibarra (1934–35, 1944–47, 1952–56, 1960–61, 1968–72) who first radicalized the emotional element common to all political speeches by constructing politics as the emotion-laden struggle between *el pueblo* and *la oligarquía*. The populist storyline that reduces complex political power relations to a simple (post-)colonial Manichean confrontation was constantly cited by ensuing populist presidents, such as Abdalá Bucaram (1996–1997), Lucio Gutiérrez (2003–2005), or Rafael Correa (2007–today). Denouncing them as 'dominant gangs, professional politicians, traffickers, and merchants of the homeland's honor' (Velasco Ibarra), 'the cause of all evil' (Bucaram) or '*mafiosos*, liars and dinosaurs headed for extinction' (Correa) (de la Torre 2010: 49, 81,181), they associate their opponents with colonial elites. This storyline constructs the very binary between the oligarchy and *el pueblo* it specifies. By generating *rabia* in the electorate against political and economic opponents through recalling past memories, voters are supposed to be mobilized.

While this Manichean rhetoric has been cited and disseminated throughout the past decades, the signifiers of *el pueblo* and *la oligarquía* were filled at different times with different meanings. Under Ecuador's first populist president, Velasco Ibarra, the Manichean rhetoric was discursively constructed as a moral and religious fight between the virtuous *pueblo* and the sinful, corrupt liberal oligarchy. It is important to keep in mind that *el pueblo* at this time included only people who had the right to vote. For Velasco Ibarra, the people – his famous *chusma* (rabble) – represented the literate lower *mestizo* classes, but excluded illiterate (mainly indigenous and Afro-Ecuadorian) men and women who were not eligible to vote until 1979, when the literacy requirement was abolished (Sosa-Buchholz 2010: 48).

When Abdalá Bucaram, who campaigned in the 1980s after universal suffrage was introduced by the new constitution of 1979, referred to *el pueblo* in his political speeches, the newly enfranchised illiterates were included. The illiterate people were transformed into political subjects through the new constitution, and Bucaram's speech acts portrayed the poor and illiterate as the incarnation of the authentic (*mestizo*) Ecuadorian nation to come under his leadership. At the same time, he portrayed the oligarchies (mainly his political opponents) as '*vendepatrias*' (selling out the country) (Freidenberg 2007: 154). The name '*vendepatrias*' recalled, through a process of citation, past memories of the exploitation of the country's (natural) resources by the Spanish colonizers. By using this insulting nickname, his electoral campaign was framed within geopolitical narratives of colonialism, aimed at provoking *rabia* against social injustice and exploitation associated with colonial times.

Rafael Correa further dramatized this populist Manichean rhetoric. His political campaign responded to people's claim *'que se vayan todos'* (that they all leave political office) during the uprisings in 2005 that brought Rafael Correa to power. Correa transformed the people's claim *'que se vayan todos'* into a performative speech act when dissolving the National Congress and convening a Constituent Assembly. He initiated his electoral campaign in 2006 with a clear anti-establishment message against the *poderes fácticos* (de facto powers), the *'enemigos de la Patria'*, 'those who want to safeguard their privileges' such as traditional politicians and parties, economic elites and the 'seditious press' (cited in Zepada 2010: 178, 179, 180). His opponent Gustavo Noboa was discredited as an exploitative capitalist oligarch who would run the country as if it were one of his own banana plantations (Conaghan and de la Torre 2008). While defining *'la oligarquía'* and *'los patriotas'* in a slightly different way to his populist predecessors, Correa's rhetoric and the (post-)colonial vocabulary cites a similar colonial and post-colonial storyline to that employed by his predecessors in their speeches. He reiterates the same populist storyline when portraying politics as the struggle between the oppressed *pueblo* and (post-)colonial elites who, similar to the Spanish colonizers, exploit the country and its people for their own benefit. As Zepada (2010) shows, Correa's campaign slogan *'la patria ya es de todos'* (the patria now belongs to everybody) implicates that *la patria*, just like during colonial times, is the property of *'dueños particulares'* (a few landlords). Comparing himself with liberator Simón Bolívar, his political speeches reduce the complex economic and political power relations of a globalized world to a simple plot. Equating his political opponents with the colonial elites, his rhetoric aims to mobilize the constituents through emotions of *rabia* associated with the colonial encounter.

AMOR FOR THE PEOPLE, AMOR FOR THE POPULIST LEADER

'Populist politics has always been about passion. It has been passionate in the romantic sense. It has been passionate in the sense of strongly emotional, the politics of personal charisma rather than the politics of abstract policy. And it has been passionate in generating strong feelings of love for the people, and hatred for those who are defined as outsiders' (Kampwirth 2010: 1).

As the quote highlights, the emotion that complements the Manichean rhetoric of *rabia* against opponents is about the *amor* of the populist leaders for their *pueblo*. The emotional performance of love is most prominently staged in Ecuadorian politics by President Abdalá Bucaram (1996-1997). In speech acts that were accompanied by close body contact with his supporters and especially their children, Bucaram performatively constructed himself as father of the poor. De la Torre (2010: 93), who carried out an ethnographic study of Bucaram's political rallies in

1996, points out that Bucaram constantly emphasized in his campaign that 'he loved *el pueblo,* he loved the poor, he loved Ecuador' (de la Torre 2010: 95). Contrary to the widespread assumption that emotions are inherently female, the example of Bucaram shows that it is dominantly men who emotionalize populist political spaces. Populist presidents 'describe their feelings for *el pueblo* in very emotional terms' (Kampwirth 2010: 2). The populists' performances of passion and love for the people that are linked to ideas of fatherhood and patronage, gender the political spaces in a specific way.

Bucaram further genders the political space through his performance as loving 'leader of the poor' with a religious storyline, portraying himself as Jesus Christ who 'sacrifices [himself] for the poor' (de la Torre 2010: 93). De la Torre (2010: 93) states that

> 'Bucaram imitated the televangelist style of praising the Lord with music, songs [...]. This is why he jumped off the platform after each speech and walked through the masses. The audience tried to touch their leader, who, like Christ, touched the people to heal and redeem them'.

Video 5: Campaign Spot Bucaram: http://www.youtube.com/watch?v=N5egRz2RJW8

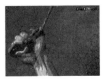

Figure 21: Bucaram 'bathing' in the crowd   *Figure 22: Bucaram crucified*   *Figure 23: Bucaram crucified*

Bucaram's performance follows a religious script. Equating the devotional love of Christ for his followers with the *amor* between himself and his supporters, Bucaram emotionalizes the spaces of campaigning in a spiritual way. Closeness to his supporters was further generated by emphasizing in his speeches and through his way of living that he belonged – just like Christ who was born in a stable – to *el pueblo*. '[H]is way of speaking; his penchant for *guayaberas* (shirts) and jeans, his passion for playing soccer, his way of eating with a spoon like the poor, rather than a fork and knife like the rich, and his love for popular Ecuadorian cooking' (de la Torre 2010: 91) performatively (re-)produced his claim to be one of *el pueblo*. Other populist presidents like Lucio Gutiérrez or Rafael Correa subsequently reiterated similar storylines to emphasize their humble origins. The storyline of the humble but loving father aims to provoke emotions in the electorate associated with the family: intimacy, trust, confidence, love, admiration etc. It is the general mistrust in political elites – generated through memories of exploitation and mistreatment by (post-)colonial elites – that has made people fall for political 'outsiders', as the following quote highlights: 'He [Lucio Gutiérrez] was born of humble origins, and I am also from humble origins ... I trusted Gutiérrez' (Reel 2005: A13).

De la Torre (2010: 97) shows, however, that just like in a family, poor people's love towards *their* politician is closely linked to their need of protection and assistance. 'The vote works as the poor's credit card' (Vilas 1998: 132) which provides access to social services. In the campaign, the poor and excluded exchange their feelings of admiration and loyalty for the populist leader for access to economic resources and services, as a militant supporter of Bucaram states:

> 'I love my party. [...] My heart tells me to be *roldista* [party of Bucaram] and I will be one until God calls me back home. Why? Because it is a very humble party, all, all of them, because of the President... he has helped all of us to find a job. He has given us a job. This is why people join the party' (cited in Freidenberg 2007: 155, own translation).

The populist script of campaigning assigns clear rules and roles to each of the participants in the campaign: The populist politician comes to the poor *barrios* bearing gifts and promises. The poor promise to vote for the candidate in exchange for his assistance in improving their life situation.

EMOTIONAL GEOGRAPHIES OF LOCAL CAMPAIGNING

Analysis of the campaign practices of Ecuador's most famous populist politicians has shown that the Manichean rhetoric employed in their campaigns generates two complementary emotions: *rabia* against the opponents and *amor* for their *pueblo*. These two emotions have become performative through constant circulation and reiteration. By recalling memories of the exploitation of the Ecuadorian *pueblo* and referring to religious and paternal feelings of devotion and love, the emotional performances align certain emotions with certain bodies: The speech and body acts assign *rabia*, contempt, anger and distrust onto the white/*mestizo* bodies of (post-)colonial economic and political elites; *amor*, admiration, passion and affection, in contrast, are assigned to the bodies of the humble, poor people and their populist leaders.

New political subjects have entered the political arena with a promise to challenge the clientelistic and populist practices of contemporary Ecuadorian politics by developing more participatory and inclusive ways of doing politics (Cañete 2004b, Van Cott 2008). As electoral campaigns are probably those political events with the greatest public exposure, the canvassing of new political subjects is an interesting site to study imitations of and deviations from national leaders' campaign performances.

RABIA: 'LOS PERROS DEL GOBIERNO'

> '*Compañeros*, it is true that we have had [political] authorities and sure they are here, watching the speech of a *kichwa* man from the Amazon. Learn your lesson and shout out loud! When they were authorities – I am referring to the "dogs" of the government, who must be

here already – when they were authorities, what did they do? They turned their back on our brothers of Dayuma and they garroted them. But there was a woman, [...] who turned in her body for the *compañeros*' (electoral speech, Marco Santi, candidate for the National Assembly, 21.04.2009, Video 6: 'Marco Santi', http://youtu.be/4ShQZcjhmsE).

Figure 24: Marco Santi
Source: Author

Figure 25: Crowd cheering
Source: Author

Listening to Marco Santi's electoral speech in which he disparages the opponents as *'perros del gobierno'* (dogs of the government), observing his gestures and sensing the emotional climate that his speech provokes in the crowd, one is reminded of the way populist leaders like Rafael Correa or Velasco Ibarra have evoked feelings of resentment by drawing on Manichean rhetoric. His speech reiterates the populist Manichean storyline that constructs politics as the emotion-laden struggle between *el pueblo* and the authorities. The emotion of *rabia*, however, is not 'stuck' to the white and *mestizo* bodies of economic and political elites, as the populist rhetoric would have it, but to the bodies of *mestizo* populists like President Rafael Correa and *mestizo* local politicians of his party *Alianza País*. At the same time, *el pueblo* is named in racialized terms by the speech act *compañeros* and *brothers* and generated as a local and indigenous community in this act of naming. Recalling the shared memories of the past event 'Dayuma' creates the community of the *Pachakutik* supporters. In the context of the campaign in Orellana, it is not necessary to amplify this word with a description or history. Marco Santi assumes that the word 'Dayuma' is enough to evoke feelings of injustice, exploitation, and suffering in the audience. 'Dayuma' stands for the detention of (indigenous) workers and the Prefect Llori on behalf of President Correa during protests in the indigenous community of Dayuma (Comisión Veedora 2008). Street riots – which since the 1990s have become a frequent means of fighting for local claims in the province – escalated in the blockade of petroleum extraction. While the other twenty-two protesters were released after a few days, Llori was detained in prison for over nine months without any clear evidence of her guilt (Human Rights Foundation, 24.09.2008).

Marco Santi's speech re-narrates this event in a specific way by 'sticking' different emotions to the differently racialized and gendered bodies involved in this incident. When calling Correa and the local representatives of his party 'dogs of the government', their bodies are constructed as dangerous, menacing but also controlling and observing bodies. By accompanying these words with a gesture that points at some (imaginary) people observing the campaign event at the back of the crowd, Santi evokes a (post-) colonial imaginary that pictures the indigenous people as surrounded, menaced and constantly observed by the colonial *patrón*. Evoking the image of dogs that watch a flock of sheep, an analogy is established to the colonial machinery of surveillance. Colonial memories of exploitation, violence and abuse are further recalled by highlighting that these dogs of the government have garroted the brothers of Dayuma. In so doing, the (post-) colonial storyline of populist Manichean rhetoric is reiterated, but the roles of the oppressor and the oppressed are aligned in a different way: The populist politicians that picture themselves as one of the *pueblo* are assigned the role of the colonial exploiters. The oppressed are the brothers (of the *Kiwcha* man) who share the same history of exploitation, injustice and neglect. Colonial missions mark the particular history of this pueblo, the extraction of valuable wood and later of petroleum. This history has generated particular feelings in the people, such as *rabia* at the constant neglect of the people's suffering by traditional political elites, feelings of hurt pride that indigenous people still have when remembering how *mestizo* political elites discriminated against them, and feelings of powerlessness against the large oil companies (informal conversation with Marco Santi). In short:

> 'Orellana is "*un pueblo*" [in the double sense of people and village] blessed by God and damned by the [local and national] governments who exploited our territory and people in an irrational way' (interview Guadalupe Llori, 17.02.2010).

The image of Orellana as an exploited *pueblo* is generated through these kinds of constantly circulating references of Orellana's colonial past and by linking its present history to this past. Marco Santi's speech introduces a third role in the postcolonial story of exploitation. In his story, 'a woman, [...] who turned in her body for the *compañeros*', rescues the oppressed (indigenous) brothers. Similar to the religious storyline of Bucaram who comes to release the humble people from suffering and poverty, Prefect Guadalupe Llori is pictured through this speech act as a martyr or as Christ-like redeemer. While in the campaign stories of Bucaram or Correa it is their masculine bodies that rescue and assist the poor, the body of a woman 'saves' the brothers of Dayuma. Feelings of hope, confidence and gratitude are associated with the body of a woman through the speech act of Santi.

The speeches of Marco Santi and Guadalupe Llori (see introduction) aim to move their audience emotionally by recalling memories related to three fundamental storylines: the Manichean struggle between the good *pueblo* and the bad populists, the (post-)colonial encounter, and the redemption of the people from their suffering through a martyr. By so doing, they (re-)produce the storylines on which populist campaigns are built. A more detailed analysis shows, however, that the emotions generated through citing these three storylines are 'stuck' to bodies that are gendered and racialized in a different way than in stories told in populist campaigns. The emotions transmitted through their performances and the emotional reactions of the *Pachakutik* supporters present in this campaign event are thus profoundly embedded in the place of Orellana and its specific geographical and historical context.

AMOR: 'AMO A MI PUEBLO'

'We want, *compañeros*, we want to serve the way we have served for the past four years, with lots of love, with lots of spirit, and especially with lots of feeling in favor of the great majorities of our *pueblo* of Orellana' (electoral speech Anita Rivas, Mayor, 21.04.2009,Video 7: 'Anita Rivas', http://youtu.be/4ShQZcjhmsE).

Figure 26: Anita Rivas
Source: Author

While dismissing the government and its local politicians is one side of the populist emotional register, the other side consists of establishing an affective bond with the people. Mayor Anita Rivas' promise that she wants to serve the people with love is very familiar as this claim follows the populist campaigning script. Nearly all *Pachakutik* candidates reiterate the promise that they truly 'love' the *pueblo* of Orellana and that they will serve the *pueblo* with abundant love. One storyline that brings the feeling of love to the center in populist campaigns is the narration about one's humble origin in *el pueblo*.

All the local candidates in Orellana draw on this storyline in one way or another when highlighting that they are 'women and men from our *pueblo*' (Guadalupe Llori), 'a *kichwa* Amazonian brother' (Marco Santi), 'a daughter of the Canton' (Maggali Orellana), or 'the mother of Coca' (Anita Rivas). Emphasizing that they have grown up in the province of Orellana, they show that they are true and authentic representatives of *el pueblo*. At the same time, through this kind of statement, they distance themselves discursively from the candidates of other political parties that 'are here only for a year or two', from the *Waira pamushkas* who come out of nowhere (speech Guadalupe Llori). It is the specific history of the province as the country's main hub for petroleum business and the shared experience of having lived in this *'pueblo de nadie'* ('nobody's land') that makes their local identity decisive for establishing a strong emotional bond with the audience:

'This *pueblo* was totally neglected. Orellana was like a *nido de lodo* [hole full of mud], totally neglected by all governments. It was a floating *pueblo*, the people just came to work in the [oil] companies, [...] , there was no identity, nobody believed that this *pueblo* could be organized, that this *pueblo* would be one day what it is now ... this was a *pueblo de nadie*' (interview Rocio Veloz, 04.02.2009).

There is a strong narrative that it was Prefect Guadalupe Llori – who previously served as Mayor – and the current Mayor Anita Rivas who turned around the destiny of the province. This narration somehow (re-)produces the image of Guadalupe Llori – and Anita Rivas to a lesser extent – as the Christ-like redeemers of this 'lost Amazon *pueblo*'. In fact, the transformation of the *'nido de lodo'* into a town is the result of performative speech acts articulated by the two women. The speech acts that promised infrastructure projects and the creation of a proper identity became performative when such improvements as canalization and asphalting, the festival of Provincialization, the beauty contest *'Nusta Kallary Runakpunapak'*, or the *'mesa de las nacionalidades'* (round table for ethnic issues), became a reality in the province.

The hard work of the two women to realize these numerous beneficial projects is considered both by the women themselves and by many supporters as proof of their love. Further, the constant reiteration of speech acts, such as 'I serve the people with lots of love', 'I love and serve the people of Coca just like my own children' (interview Anita Rivas, 25.02.2009) or 'I love my country, I love my *tierra*, I love my *pueblo* and what I do, I do with love' (interview Guadalupe Llori, 17.02.2010) performatively reassure their endless love and the sacrifice they are willing to make for their *pueblo*. As the quote demonstrates, by calling themselves 'mother of *el pueblo*', they construct themselves, in an analogy to the populist father figures, as loving mothers. Through this kind of speech act, the emotional performances of *amor* are gendered differently from those of masculine populists who emphasize their fatherly love.

The emotional geographies of love brought into being through the performances of the local *Pachakutik* candidates, however, are not only gendered, but also racialized. While drawing on the same storyline about a shared humble origin so often cited in populist campaigning, the speech and body acts of the local candidates in Orellana often emphasize their affective bond to the indigenous community. Although all candidates run for the party *Pachakutik* that is well known for its close link to the indigenous movement, the only one able to address the audience in *kichwa*, the indigenous language, is Marco Santi. His knowledge of *kichwa* and his provenance from a *kichwa* community seem sufficient to provoke passionate support from the *kichwa* constituents. Nearly all indigenous people start cheering when he welcomes the crowd in *kichwa*. Candidates who are not able to speak *kiwcha* compensate for the lack of biographical roots in an indigenous community by appearing decked out with accessories in the rainbow colors of the indigenous *Wiphala* flag. They express performatively their emotional commitment to the indigenous people by wearing indigenous symbols and emphasizing in their speeches their solidarity with the indigenous people (see speech of Guadalupe Llori in the introduction).

These examples show that the local *Pachakutik* candidates (re-)produce populist storylines and scripts in an attempt to demonstrate to their audience that they truly love them. They reiterate populist storylines when emphasizing their origin in a humble (indigenous) family of Orellana and they copy populist canvassing rituals when seeking close body contact to their supporters. While their performances are far from inventing new campaign practices or generating different emotional atmospheres, the presence of their feminine and indigenous bodies, the use of *kichwa* and indigenous symbols, and the political projects realized, serve to challenge populist politics. They do so by acknowledging the political subjectivity of social groups, such as indigenous people and women, which up until today are often not taken seriously as political citizens. The presence of indigenous bodies on the electoral stage, the welcoming words in *kichwa*, or the inviting gesture of a woman politician are the first steps in opening up the racialized and gendered boundaries of (post-)colonial spaces of electoral politics. The initially invisible effect of the new political subjects' engagement in electoral politics gradually becomes visible, for example in the strong presence of women and indigenous people, not only in *Pachakutik's* electoral canvassing in the Province of Orellana, but also in the participative assemblies of the municipality and the provincial council.

TOWARDS AN EMOTIONAL ELECTORAL GEOGRAPHY

The empirical analysis in this chapter aimed at contributing to debates in (feminist) political geography about identity, politics and social change in the Andes (Andolina, Laurie, and Radcliffe 2009, Laurie, Andolina, and Radcliffe 2003, 2005, Radcliffe 2008a, 2011, Radcliffe, Laurie, and Andolina 2002, Radcliffe, Laurie, and Andolina 2003, Radcliffe and Pequeño 2010). Focusing on emotions as a conjunctive element between new and hegemonic political subjects, it has introduced questions about emotions and politics to this body of literature. Taking into account the politicization of indigenous subjectivities throughout the Andes (Lucero 2008, Oslender 2008, Radcliffe, Laurie, and Andolina 2002, Van Cott 2008, Yashar 2006b) and the introduction of gender quotas and concomitantly the significant increase of women in electoral politics in many Latin-American countries (Araújo and García 2006, del Campo 2005, Peschard 2003, Ríos Tobar 2008), the issues addressed are relevant throughout Latin America.

On a conceptual level, this chapter has addressed the role of emotions in the constitution of political spaces and shown how a performative approach to emotions can contribute to the analysis of political change and continuity in political campaigns. I have argued that a performative understanding of emotions as developed by Ahmed (2004a, 2004b) is especially suited for analyzing emotions in contexts of political transformation as it focuses on how emotions align with certain subjectivities through the 'sticking' of specific signs to bodies; it analyses how emotions become 'sticky' through repetition; and questions how emotional performances literally 'move' both the performer and the audience. By linking Ahmed's notion of performative emotions to Hampton's (2009) work on the performativity of political speech, I have made her approach productive for emotional geographies interested in questions of (electoral) politics.

This conceptual framework can be employed to analyze the emotional dimension of processes of political change and continuity. In the first part, I have analyzed how the two emotions of *rabia* and *amor* have come to dominate the emotional atmosphere of populist campaigns. I have shown how the Manichean rhetoric aligned *rabia* to the white and masculine bodies of (post-)colonial elites, while constructing the relationships between *el pueblo* and the populist politician as full of *amor*. This Manichean rhetoric has been cited over and over again in Ecuador's populist campaigns and has therefore served to transform these two emotions into the hegemonic emotional register of campaigning. In the second part of my empirical analysis, I have questioned to what extent the so-called new political subjects merely reproduce, cite and reiterate or actually challenge the emotional register of populist campaigning. While identifying similar storylines and scripts in the local campaign under scrutiny, I have also highlighted that the emotions of *rabia* and *amor* are performed in a slightly different way as they are aligned to differently gendered and racialized bodies.

Small changes in the populist script, like the racialization of the Manichean rhetoric, the use of *kichwa*, of the *Wiphala* flag, alter the way political spaces of campaigning are constructed. The new political subjects open up the racialized and gendered boundaries of (post-)colonial populist spaces of electoral politics through the presence of their feminine and indigenous bodies and the way emotions of *amor* are explicitly 'stuck' to the bodies of women and indigenous people. Through emotionally addressing these social groups that were formerly excluded from political participation, they expand the boundaries of the political by challenging the hegemonic gendering and racialization of political spaces as white/*mestizo*, masculine and upper-class.

The performative understanding of emotions applied in this chapter emphasizes that the emotions I observed during the local campaign in Orellana can only be understood as being generated by, and expressive of, the particular historical, economic, political and social-cultural context in which the constituents live. Hence, along with Tolia-Kelly (2006), I would like to plead for the importance of embedding research on emotional geographies in theories of difference. My research on the emotional geographies of local campaigning has shown that it is crucial to take into account the (historical) power geometries that shape and facilitate specific kinds of emotions by gendered and racialized bodies due to their identity-specific historical and everyday experiences.

6.
THE INTERSECTIONALITY OF EMOTIONS IN CAMPAIGNS

> 'The point is that democracy without impassioned argument, without rhetorical flourishes, and without the sharing of laughter, anger or grief is democracy without vitality' (Hoggett 2009: 149).

Scholars from different disciplinary backgrounds have become increasingly aware that emotions matter in the constitution of the political. In 'Zorn und Zeit', the German philosopher Peter Sloterdijk (2008) examines how anger has been managed throughout political history. Arjun Appadurai (2006) in his book 'Fear of Small Numbers' highlights the conflictive emotions between majorities and oppressed minorities. Chantal Mouffe (2007) sees antagonism between different political ideologies as a source of passionate politics. Most of this scholarship emanates from emotional responses to 9/11 and the perceived threat of global terrorism (Appadurai 2006, Butler 2005, Sloterdijk 2008). In a similar way a 'new geopolitics of fear' has emerged within the sub-discipline of critical geopolitics (Gregory and Pred 2007, Megoran 2005, Pain 2009, Pain and Smith 2008, Sparke 2007), focusing on the transnational dimensions of fear and how fear is produced and imagined connectedly from one site in the world to the other. In this chapter I argue, however, that it is worth studying the emotional geographies of everyday local politics in order to understand the constitutive role of emotions for the political. As electoral campaigns can be seen as 'dramas of identity and difference' (Spencer 2007: 78) in which emotional performances play a crucial role, they are an ideal place to study the emotional geographies of the political.

Drawing on ethnographic research conducted during the electoral campaign of 2009 in Ecuador's local politics, this chapter is interested in the role, emotions play in the constitution of local political spatialities in postcolonial societies. Ecuador's recent campaigns are characterized by the increasing presence of women, indigenous and Afro-Ecuadorian people, who have been (post-)colonially excluded from political citizenship, after the introduction of a gender quota law (Vega Ugalde 2005) and the implementation of indigenous parties in the 1990s (Lluco 2005, Sánchez-Parga 2007). This chapter is especially interested in the emotional geographies of electoral campaigning these so called new political subjects' are producing. Employing the concept of intersectionality (Crenshaw 1989), the empirical research aims to understand how the (collective) emotional performances within political campaigning result from, address and challenge (post)colonial experiences of racism, patriarchalism and classism

BRIDGING EMOTIONAL AND POLITICAL GEOGRAPHIES

In recent years there has been an increasing interest in the role of emotion and affect as central to the constitution of space and place in social and cultural geography (Anderson and Smith 2001, Bondi, Davidson, and Smith 2005, Davidson and Milligan 2004, Pile 2009a, Smith et al. 2009a, Thien 2005). Given the importance of emotions for the constitution of political space, it is surprising, however, that accounts focusing on politicized emotions from a geographical perspective have been few.

I build my argument on two bodies of literature that address the political relevance of emotional geographies: emotional geographies of social movements and feminist approaches to emotions. The first strand engages with geographies of social movements that deal with the role of emotions as key elements in activism (Bosco 2004, 2006, 2007, Oslender 2007, Wright 2008). This body of work ties in with a wider range of literature around emotions in social movements (Goodwin, Jasper, and Polletta 2001, Kemper 2001, Scott 1990, Yang 2000) which addresses collective feelings and the relation between emotions, political activism and identity such as ethnic groups (Oslender 2007, West-Newman 2004), feminist movements (Holmes 2004, Taylor and Rupp 2002b), and motherhood (Bosco 2004). Bosco's (2004; 2006; 2007) research on the emotional labor of the *'Madres de Plaza de Mayo'* who protest against the human rights violation of Argentina's dictatorships by mourning publicly and collectively the loss of their sons is an interesting example that shows how social movement activism revolves around gendered identities and concomitantly gendered emotions. Similar to Bosco, most of these studies focus on identity politics, claiming that identity-specific emotions such as motherly love (Taylor and Rupp 2002b) or racialized postcolonial anger (West-Newman 2004) lie at the heart of social movement activism.

Although, as Pain (2009) highlights, feminist political geographers to date have had little to say about the role of emotion within the political, feminist geography in general has been concerned with the gendering of emotions since its beginnings by challenging binaries around emotion/reason, public/private etc. (Blunt and Rose 1994, Bondi 1990, Rose 1993). While initially the gendering of both emotion and politics was a central concern of feminist politics, recent accounts of feminist (political) geography have emphasized that

> 'it would be ethnocentric, if not racist, to assume that gender is always and everywhere the primary basis of oppression, persecution, or exclusion (Anzaldúa 1987, Mohanty 1991). Relations of class, race, caste, sexuality, religion, nationality, ethnicity, and other axes of affiliation are potentially exclusionary, discriminatory, and even violent' (Hyndman 2004: 309).

These gendered, racialized, sexualized etc. experiences of exclusion, discrimination, and violence result in and provoke emotions in both individuals and social groups. Connecting to a broader body of work on emotions within feminist and gender studies (Ahmed 2004b, Holmes 2004, Nelson 1999a, Ngai 2005), feminist geographers have addressed emotions as different as everyday fear (Valentine 1989), fear of global terrorism (Pain and Smith 2008, Radcliffe 2007), hope for change (Sparke 2007, Wright 2008), sadness and longing (Pratt 2009), etc. Feminist geographers, however, do not merely engage empirically with emotions but contribute through their work and critique to (disciplinary) debates around emotional and affectual geographies (Bondi 2005b, Bondi, Davidson, and Smith 2005, Tolia-Kelly 2006).

Of particular relevance for the present study is Tolia-Kelly's (2006) critique of the universalist thinking in and historicity lacking from Nigel Thrift's (2004) account of a 'spatial politics of affect' in particular and the neuropsychological approaches of affectual geographies in general. She argues that, due to historical and spatial power relations, racialized, gendered and sexualized bodies have different capacities to experience and affect the social space around them. Tolia-Kelly's argument serves as a conceptual starting point for my endeavor to conceptualize the emotional geographies felt, experienced, and displayed by racialized and gendered bodies in Ecuadorian political spaces. Drawing on ethnographic observations made while accompanying women who ran as candidates for local elections, I explore the utility of an intersectional perspective in the empirical study of emotional geographies.

RESEARCHING EMOTIONS FROM AN INTERSECTIONAL PERSPECTIVE

Feminists of Color (e.g. Anzaldúa 1987, Crenshaw 1989, Mohanty 1991) have developed the concept of intersectionality to show that women experience oppression differently, and to varying degrees, depending how gender intersects with race, class, sexual orientation and other categories of social organization. The concept makes it possible to 'think of race, class and gender [and age, location, sex, disability etc. *author's addition*] as different social structures individual people experience simultaneously' (Valentine 2007: 13). As intersectional research focuses on the interrelations between different systems of oppression, it is particularly suitable for looking at the postcolonial framing of emotions.

There is an intense discussion, both in gender studies and feminist geographies, how to research intersectionality. Case studies (McCall 2005) are still seen as the most feasible way of handling the complexity of intersectional research. Feminist geographers frequently focus on one specific 'site' such as the workplace (McDowell 2008) or the community (Lobo 2010). Most analysis draw on qualitative research methods. By doing so, intersectionality is researched as 'lived experience' (Hopkins and Noble 2009, Valentine 2007). Valentine's (2007) narration about Jeanette, a d/Deaf woman who experiences her deafness very differently along her life-course and within different spaces, is an example of how particular identities are given importance by individuals at particular moments and in specific contexts. While these narrations highlight the individual experience of intersectionality, structural and discursive intersectionalities are addressed only implicitly. Only few studies consider that categories like race and gender are not primarily a question of skin color and sexuality, but of imperial exploitation and subjugated labor (McClintock 1995). Hence, as Valentine (2007: 19) herself points out,

> 'the existing theorization of the concept of intersectionality overemphasizes the abilities of individuals to actively produce their own lives and underestimates how the ability to enact some identities and realities rather than others is highly contingent on the power-laden spaces in and through which our experiences are lived'.

I propose to overcome this problem by combining structural, discursive and performative approaches in this chapter. Based on Winker's and Degele's (2011) work, I develop a three-level approach to analyze the intersectionality of emotions. First, by highlighting structural inequalities in Ecuadorian society, I contextualize the collective feelings expressed by the new political subjects based on injustice and inequalities. Second, I analyze discourses, norms and ideologies on the representational level in order to show how the political discourses around rationality, masculinity and whiteness justified the exclusion of 'racialized' people and women by regarding them as too emotional. Third, by (re-)constructing the importance of emotionality in the identity performances of new political subjects, I aim to understand how these (re)-produce and challenge stereotypical imaginations of identity-specific emotions through their performances.

Structural inequalities, symbolic representations, and identity constructions are expressed through social practices (Bourdieu 1998). Ethnographic methodologies are one possible way to capture practices. The methodology applied reflects recent calls for a shift to ethnographic methodologies in (feminist) political geography (Megoran 2006, Sharp 2004a). Between 2008 and 2011, I spent a total of 18 months in the three Provinces of Esmeraldas, Chimborazo and Orellana, accompanying political women in their everyday activities and conducting interviews with female politicians. The semi-structured and partly biographical interviews with forty local female politicians, among them seven provincial candidates for the national assembly, three (vice-)prefects, three (vice)mayors, thirteen town councilors, fourteen rural councilors as well as with fifteen women heading local women's organizations. Further, I accompanied two prefects, two mayors and four councilors for several weeks in their everyday political activities and later on video-recorded their campaigning activities, spending a week in each province during the electoral campaign in 2009. Once women noticed that I had a camera, they asked me to record their activities. Photographs and videos were later used to (re-)construct a close description of events, focusing on what the candidates' bodies were doing, their facial expressions, their interactions, and their gestures, rather than asking them 'how do you feel?'

STRUCTURAL INTERSECTIONALITIES AS CAUSE AND TARGET OF COLLECTIVE FEELINGS

> Around 500 people have taken up the indigenous and *campesino* (peasant) party *Pachakutik's* invitation to the primary elections in the Coliseum of Coca, an Amazonian provincial capital. Despite the heat, people are waiting patiently until suddenly the music is turned on and a local

musician sings a folk song composed especially for Guadalupe Llori, the Prefect of the Province of Orellana. She is one of the two female prefects; all the other 22 provinces are governed by men. The eyes of the people around me sparkle while singing the campaign song with passion, euphorically waving the rainbow flags of the indigenous *Pachakutik* party, clapping their hands endlessly, shouting *'Que vivan las mujeres'*. The song is followed by a dance group of indigenous children, all dressed in traditional dress, dancing to the rhythms of *cumbia*. The first to speak is an indigenous man. He welcomes everybody in *kichwa*. As my *kichwa* is very poor I only understand a view words, when he says '*Alli punsha, Alli Shamushka ...indígenas kichwa, shuar, waorani, ... Yupaichani.*' The indigenous woman next to me explains that he is asking the people to fight together against the discriminatory practices of the government. Then an election committee is appointed and people are asked to register in the candidate list. Three men are volunteering but it seems difficult to find the same number of women – a requirement by the electoral law after the introduction of the gender quota. All candidates are then asked to present a short speech. Marco Santi refers to his ancestors and to the liberation from Spanish rule and concludes his speech with the words 'now the time has come that we, the *indígenas,* can have a say in the political world'. People shout with excitement, '*que viva Marco Santi*'. Maggali Orellana, a pregnant woman wearing ear rings in the colour of *Pachakutik,* accuses the current Prefect of discriminating against the Amazon *pueblos* (indigenous tribes) just because they defend their land. Then the microphone is passed on to an indigenous woman. Kantapari Nuni seems very nervous and insecure. She gives the shortest speech, stumbling over words as she speaks in Spanish, which is not her mother tongue. Then the voting takes place (Field notes, Coca, January 2009).

By looking from an intersectional perspective at the narration, both the persistence of 'structural intersectionalities' (Crenshaw 1994a) and ongoing transformations that are currently take place within Ecuadorian politics become evident. Taking into account that until 1979 literacy requirements effectively excluded up to 70 per cent of women (Quezada 2009: 190) and nearly all indigenous people from taking part in elections (Yashar 2005: 37), the (re-)narrated event which includes both female and indigenous candidates is a testimony of the ongoing transformations within the post-colonial political spaces regarding the extension of citizen rights to social groups that have been (post-)colonially excluded. The outcomes of the gender quota and the political success of the indigenous movement become visible through people like Marco Santi, Maggali Orellana and Kantapari Nuni.

The interplay between the consolidation of women's organizations, the support for these by international (non-)governmental agencies, the International UN decade for women and Ecuador's ratification of CEDAW (International Convention on the Elimination of All Forms of Discrimination against Women) in 1981 resulted in a long struggle for a gender quota law, finally approved in 2000 (Vega Ugalde 2005). Now, political parties are obliged to include 50 per cent women on their voting lists. This quota makes Ecuador a pioneer in the implementation of gender policies (Bedford 2008). Although the electoral outcomes are far from reaching parity, women's political participation has increased significantly on all political levels. *Mestiza* and Afro-Ecuadorian women, however, have benefitted to a greater extent from the quota than indigenous women who are underrepresented both in comparison to indigenous men and non-indigenous women (Radcliffe, Laurie, and Andolina 2002).

Indigenous mass movement protests during the 1990s argued fiercely for the state's responsibility to fix past racialized injustices. As a response to the neoliberal policies of social exclusion and to the crisis of traditional political parties, the indigenous movement formed an electorally viable political party in 1996 (Lluco 2005, Lucero 2008, Macas 2002, Martínez Novo 2009a). The indigenous party *Pachakutik* has been especially successful in municipalities, where it intends to construct a politics that contests 'the exclusion of indigenous and low-income peoples from formal politics' (Radcliffe, Laurie, and Andolina 2002: 290).

The emergence and actions of both movements, which intersect in certain moments and spaces, can be interpreted as decolonization processes, as the increasing presence of indigenous people, Afro-Ecuadorians and women within representative politics challenge the (post-)colonial domination of male and white/*mestizo* subjects. The political struggles of both movements build on the emotionalization of their political demands against structural discrimination and, concomitantly, the emotionalization of political space. The narration exemplifies how political spaces are designed in a way that invokes emotional response. The use of highly symbolical elements, such as the indigenous language *kichwa*, traditional cloths, and the rainbow-colored *wipala* flag, which is a symbol from the Incan Empire, along with the increasing use of traditional music and dance performances, reconstruct the political spaces in order to manage and instrumentalize collective feelings. The ethnographic narration evidences how 'collective emotions' (Kemper 2001: 59) result from and address (post)colonial structural intersectionalities. The anger against the long neglect of the Amazonian provinces mirrors (post-)colonial feelings about racialized economic inequalities, the rage against the exclusion of indigenous people from electoral political space dates back to colonial ideas of citizenship, or indigenous women's fears of speaking in public, which result from structural inequalities.

Intersectionality becomes evident in the case of indigenous women like Kantapari Nuni, who not only have to overcome their anxiety about speaking in public but also the challenge of speaking in a language which is often not their mother tongue. Gendered and racialized inequalities regarding years of schooling and a quality of education which differ significantly between girls and boys and between bilingual rural (*kichwa*) and urban (*mestizo*) schools can be seen as a cause of these anxieties (Garcia-Aracil and Winter 2006, Martínez Novo 2009b). A gendered and racialized lack of voice has been intimately linked to social injustice, economic inequality and political disempowerment throughout Ecuador's (post-)colonial history. As Howard (2010) highlights for Bolivia, even when indigenous politicians use Spanish, they suffer from a lack of access to the dominant discourse and a perceived lack of competence. As the quote exemplifies, however, indigenous leaders feel increasingly confident and even proud to use *kichwa* in their public speeches. The use of *kichwa* and the emotions that accompany it are examples of how emotions that were brought into being by structural inequalities can be converted in emotions of resistance and subversion. Thus, the political spaces that are brought into being by these emotions contest the hegemonic discourses which define the political as white, *mestizo*, masculine, and rational.

DISCOURSES AROUND THE WHITENESS AND MASCULINITY OF ECUADOR'S POLITICAL SPACES

Imaginaries of the political in Ecuador have been dominated (post-)colonially by masculine and white superiority (Radcliffe and Westwood 1996), enlightened ideas of masculine rationality, and a gendered and racialized public-private divide (Mills 1997, Pateman 1988). Women's, indigenous and enslaved afro-descendant people's emotionality was marked as potential danger for political judgment. Discourses of masculine and white supremacy resulted in and were at the same time (re-)produced through everyday political practices like an exclusionary electoral law or an indigenous tribute (Prieto 2004). Despite the gender quota law and the increasing presence of indigenous leaders, political spaces continue to be imagined discursively as white/*mestizo* and masculine and concomitantly as spaces in which women and indigenous people are literally 'out of place':

> 'It is that ... always politics, all the life, politics have always been done by men – everything – be it organizations of the *comunidades*, be it the *barrios*, be it in the political parties, everything is ruled by men. They told us that we are not suited to be politicians as we are too hysteric, moody, and affective' (interview town councilor, March 2010).

(Post-)colonial hierarchies between emotion=private and thought/reason=public relegated 'being emotional' (Ahmed 2004b: 4) to certain gendered and racialized bodies. Consequently the social movement struggles of women and indigenous people in Ecuador have frequently been dismissed as too emotional and therefore have not been taken seriously by local and national governments. These discursive binaries are blurred by men's everyday emotional performances on the political stage e.g. during populist speeches where they fiercely insult the opposition and shout enthusiastically political slogans to cheer on their audiences (de la Torre 2000). While dichotomous discourses around the masculinity of rationality and femininity of emotionality construct the masculine citizen as emotion-less, it is hegemonic masculine political subjects who actually define which emotions are considered appropriate and which are dismissed as being *too* emotional, *too* feminine in daily political life. A similar argument can be made for degrading indigenous political performances as primitive, not sufficient sophisticated and hence never able to meet the masculine and white/*mestizo* norms as an indigenous women representative points out in an interview:

> 'They say to us "why do the *indios* want to be something superior if they don't even know how to give a proper political speech, not even professionals will be able to do anything, they will just be there to adorn the picture of the candidate"' (Interview indigenous woman representative, February 2009).

The powerfulness of discourses around the whiteness and masculinity of politics becomes evident when analyzing these quotes from an intersectional perspective and taking into account that they are narrated by women who self-identify as indigenous. The first quotation highlights that the constitution of exclusively indigenous political spaces such as the community organizations also base on ideas of masculine superiority. The second quotation relates that indigenous leaders still are considered 'out of place' by many *mestizos*. Hence, indigenous women are discriminated 'first for being a woman and second because of their ethnicity' (Interview mayor, February 2008). Due to gendered and racialized processes of Othering, indigenous women do not fit with the predominant (post-)colonial imaginations about a rationalized, emotionless politician who 'has to have blue eyes, blond hair, a European stature' (ibid.). Thus, indigenous women are 'hit' – to speak in Crenshaw's (1989) words – by the interlocking systems of racism and patriarchalism, facing both a 'patriarchal and mono-ethnic political system', as the first indigenous Secretary of Foreign Affairs Nina Pacari (2005: 73) states. Their political exclusion frequently is justified either by generalizing racist, gendered and classed arguments around (indigenous) women's lack of education and professional capacities (and thus lack of rational thinking) or by the indigenous complementary gender order which scripts women as responsible for the reproduction of indigenous culture, including the emotional work within the family and community. Indigenous and *mestizo* women alike, however, have deployed maternalist discourses around women's 'natural' emotionality strategically (Herrera forthcoming). Women's increasing political participation was framed by a deep political crisis during the 1990s, consisting of mistrust in political institutions and various presidential coups (Cañete 2000, Rivera Vélez and Ramírez Gallegos 2005). Women justified their political participation by their moral and emotional superiority and in consequence their ability to literally 'clean up' the masculine 'dirty', corrupt spaces.

> 'Idealizing motherhood, women are encouraged to deny their own interests [...]. This suffering for others is often interpreted as women being more able to "feel" the needs of the community. Motherhood also stands for ideas of self-abnegation and rejection of self-interests' (Craske 1999: 4).

This strategic reframing of women's emotionality through the women movement is an example how (post-)colonial discourses around the rationality of masculine and *mestizo* political subjects are challenged. As a result, the emotional geographies of political spaces are constantly negotiated and (post-)colonial exclusions of certain feminized and racialized bodies from the spaces of politics become increasingly contested.

EMOTIONAL PERFORMANCES AS PART OF POLITICAL IDENTITY CONSTRUCTIONS

A performative approach to the intersectionality of emotional geographies offers the possibility to focus on the *doing* of emotions (Ng and Kidder 2010). By demonstrating how politicians emphasize certain gendered, racialized or classed emotions, according to the audience they seek to address and the political claim they aim to make, I illustrate how they challenge essentialist identity politics by blurring any fixed identity. I exemplify my argument by drawing on observations made while accompanying Mary Mosquera, Vice-Major of Esmeraldas in her campaign in a poor suburban neighborhood in Coastal Esmeraldas.

> Once we get to the *barrio popular*, Mary, the Vice-Major arrives in another pick-up together with her son. People start with an electoral choir, dancing along the rhythms of '*vota 15, vota Mary*'. The *barrio* with its sandy roads and small wooden houses had felt pretty silent and unpopulated when we first arrived, but suddenly is alive as the people from the campaign team with their blue and orange shirts and baseball caps are all over the place. In the first house, Mary is confronted with a propaganda poster from the government party *Alianza País*, which campaigns for a 'new socialism'. When two women open the door upon her knock, Mary criticizes fiercely the falsity of the government and shouts with insistence while her body gestures and especially her fist emphasizes her anger: 'I was born with the revolutionary political organization, and you should know that the *movimiento popular democrático* is the only *real* left-wing party, the only *real* representative of the *sectores populares*. We as poor women, working women, peasant women, teachers, we can't be of any other class, you should not vote for those racist *pelucones* of the government who call us "*negros de mierda*"' (Field notes, Esmeraldas, April 2009).

While Mary throughout the day and in every interaction speaks in a very loud and energetic voice, in the last part of her speech she raises her voice even more. Her anger and disgust about the incident a few days ago when President Correa called the Afro-Ecuadorian Major '*negro de mierda*' have reactivated (post-)colonial feelings of discrimination, embedded in a post-colonial discourse of white/*mestizo* supremacy. In the light of these verbal attacks, Mary positions herself as black, a marker that she does not draw on in other moments. In an interview a few months later she states 'I am *negro*, but I am not *negrista*' which means that she *does* identify as Afro-Ecuadorian but does not favor doing identity politics on this racialized identity. Instead, she denies racial difference in support of class unity. Mary's denial of ethnic signification has to be understood as part of her party's socialist rhetoric of a unified (race-less) proletariat. Sloterdijk (2008) shows that the communist 'anger economy' created a collective proletarian group by managing their collective rage against economic elites. In positioning herself as socialist, she draws on these collective feelings managed by the communist 'World Bank of anger' (ibid. 170). By doing so, she constructs her political identity as a class identity in favor of the poor people. As the household she addresses can be considered poor, her emotional reference to her poor childhood is an attempt to position herself as representative of the *sectores populares* and evoke common feelings of rage against institutionalized class-specific economic injustice.

Mary further stresses in her interactions with women emotions of woman- and motherhood. Drawing on maternal feminists' assumptions of difference (Sapiro 1981), she claims that women have particular feelings, necessities and interests which can only be represented by women:

> 'It is important that there are women in the *municipio* who respect us, women who feel the *dolor del pueblo*, the pain of her sister, women who feel women's necessities' (Field notes, Esmeraldas, April 2009).

West (1999) has shown that women's experience of violence, injury and discrimination and the resulting emotions have been crucial to feminist politics. Although

Mary does not identify herself as 'feminist', she draws on feminist strategies, 'which have mobilized around the injustice of that violence and the political and ethical demand for reparation' (Ahmed 2004b: 172). Further, relating to the same argument of maternalist identity politics she stresses that women prioritize a political agenda that favors women's concerns such as health and schooling and in general 'do' politics differently due to their gendered emotionality:

> 'We as women think in two ways: with the political logic, with the professional logic, but we also think with the mind of a mother, with the heart of a woman, for example that there is not only a park but also flowers, because it is the flowers that decorate the park' (Field notes, Esmeraldas, April 2009).

As Mary's example illustrates, bonding emotions are intended to be invoked in the electorates by displaying an 'intersectional consciousness' (Frederick 2010): First in regard to race, by hinting at racist practices of the national government party and hence discourses around white/*mestizo* superiority; second in regard to class by declaring oneself as having grown up in the same circumstances and thus evoking collective feelings against structural inequalities and third in regard to gender by emphasizing the solidarity of woman. In the course of the day she (re-)constructs her identity in manifold ways. Each situational identity construction is accompanied by a set of emotions and feelings related to her own or the voter's specific lived experiences. The displayed emotions frequently mingled their diverse sources like colonial and post-colonial racist prejudices against black people, simultaneously violent and joyful experiences of being a woman, socialist rhetoric of a united proletariat and experiences of poverty. By doing so, the emotional performances splatter the strict boundaries of fixed identity categories, establishing temporary and spatially unique embodied relations between Mary and the people she talks to.

THINKING EMOTIONS FROM AN INTERSECTIONAL PERSPECTIVE

> 'Without passions, sentiments, and their affects there would be no (political, *added by author*) geography worth the name' (Smith et al. 2009b: 10).

On a conceptual level, this chapter has set out to expand feminist approaches to emotional geographies by integrating the concept of intersectionality. By doing so, I have underscored the importance of Tolia-Kelly's (2006: 214) call for an emotional geography 'sensitive to power geometries'. The empirical case study aimed to translate her theoretical claim into research practice, showing how present power relations and their corresponding emotions within Ecuadorian society are linked to histories of (post-)colonialism, domination, subordination and resistance. Political spaces continue to be shaped by racialized, gendered and social hierarchies. In consequence, their construction is influenced by the way in which marked bodies experience and express emotions within political space. An intersectional perspective offers the opportunity to research the entanglements and relations between collective feelings and identities. By showing the entanglement of

(1) structural inequalities which lead to different capacities to experience and affect the political emotionalized space around them, (2) gendered, racialized, and classed discourses which constitute the political space, and (3) localized emotional performances which address culturally meaningful emotions, I plead for the importance to take the reciprocity between structure, discourse, and performativity into account when studying emotional geographies.

As the empirical data suggests, an intersectional analysis of emotional geographies needs to be attentive to how identity categories and power relations become relevant in emotional performances and how these categories intersect with each other. The empirical account challenges the proposed and often criticized stability and essentialism of identity politics (Noble 2009) by showing how the candidates in one moment place a certain identity category in the foreground through their emotional expressions, neglecting it altogether in another moment. By asking why exactly these intersections become crucial for the emotional performances on specific political stages, a better understanding of the relation between identity, emotion and politics can be developed.

INTERSECTIONALITIES

7.
RETHINKING GENDER QUOTAS INTERSECTIONALLY

'It is not so much that women have a different point of view in politics as they give a different emphasis. And this is vastly important, for politics is so largely a matter of emphasis' (Crystal Eastman, 1925).

ACCESSING ECUADORIAN POLITICS

'All persons are equal and shall enjoy the same rights, duties and opportunities. No one shall be discriminated against for reasons of ethnic belonging, place of birth, age, sex, gender identity, cultural identity, civil status, language, religion, ideology, political affiliation, [...]' (República del Ecuador 2008: Art. 10, 2).

'Yes, we women really have gained a space, and we have to keep on fighting for this space. The old politicians [masculine form] have lost their space, because now they have to incorporate women because of the quota law' (interview town councilor Martha Noboa, 12.08.2008).

'I think the gender quota is an important mechanism as it guarantees the access [to electoral politics]. The problem is that it does not guarantee the participation of ethnic minority women. Revising the conformation of the party lists, you find mainly *mestiza* women. The electoral participation of indigenous and Afro-Ecuadorian women is still minimal' (interview *Asambleista* Alexandra Ocles, 3.10.2010).

Significant transformations have occurred in Ecuador, as new social movements have called for more inclusive spaces of democracy during the last two decades (Radcliffe 2000, 2008b, 2011, Van Cott 2008, Yashar 2006b). Women, indigenous, Afro-Ecuadorian, and *campesino* movements have claimed for the expansion of democratic citizenship towards social groups who were post-colonially excluded from the access to electoral politics. Their struggles have resulted in a number of affirmative measures that seek to open up the (post-)colonial boundaries of institutionalized politics. While indigenous people have gained access to electoral politics through the foundation of ethnic parties (Yashar 2006b), a gender quota law has fostered the presence of women in electoral politics (Vega Ugalde 2005). Being introduced in 1998, by now Ecuador's quota law is considered one of the 'most advanced quota laws' (Peschard 2003: 25). The electoral law requires 'the obligation to comply with the principles of equity, parity and alternation between men and women on [both the principal and alternate] multi-persons candidate lists' (República del Ecuador 2009: Art. 3).

As the three introductory quotes aim to highlight, while Ecuador's Constitution addresses multiple forms of discrimination, the quota law focuses in a unitary way on one single category, namely gender. The (post-)colonial polity, however, has been built on interlocking systems of exclusion based on gender, ethnicity, race, class, sexuality and locality (Prieto 2004, Prieto and Goetschel 2008). Due to the structural intersectionality that defines access to electoral politics, women have benefitted to different degrees from the quota according to their ethnic, class and local positioning as Alexandra Ocles, the first Afro-Ecuadorian female national deputy, highlights in the third quote.

This chapter takes up the task of evaluating the impact of the gender quota on the way local spaces of politics are constituted from an intersectional perspective. The first section of the chapter briefly reviews some of the literature on women and electoral politics in order to show the importance of introducing intersectional thinking in electoral geography. The second section introduces the methodology used in the empirical study on women politicians in Ecuadorian local politics. The third section discusses the results of local elections in Ecuador from an intersectional perspective to show the effects of the quota on descriptive representation. The forth section further engages with questions of substantive representation along three case study localities. The conclusion discusses geography's contribution to the study of the effect of affirmative measures on the constitution of spaces of institutionalized politics.

INTERSECTIONALITY IN ELECTORAL POLITICS

'In a material sense, if people simply cannot get to a place where issues are deliberated – whether because they lack transportation, are mobility impaired, they simply do not know that an issue is being deliberated, or they are denied access to deliberative venues because they lack formal status as a citizen – then the polity is similarly unrepresentative of the people who live in the territory associated with it' (Staeheli et al. 2002: 994).

Reading the quote from Staeheli et al. against the backdrop of the (post-)colonial history of institutionalized politics in the Andes, the question who has been denied access to deliberative venues on what grounds is crucial. While women have formally gained suffrage already in 1929, literacy requirements to vote, which were only eliminated in the constitution of 1979, basically denied nearly half of Ecuador's population the access to political citizenship (Prieto and Goetschel 2008). Due to the unequal access to education, women, indigenous and Afro-Ecuadorian people as well as rural and underclass populations had no right to enter the spaces of institutionalized politics both in a metaphorical sense as they were not allowed to vote and run for office but also literally, as frequently they were often denied access to the buildings of the municipality or the congress. Further, as the spaces of electoral politics such as municipalities were situated in urban centers, rural dwellers often had to shoulder long journeys and hours of waiting to present their concerns to mainly white or *mestizo*, male, urban politicians. As consequence, the (post-)colonial polity was unrepresentative of the people who lived in the colonially marked territories associated with it. Hence, Ecuador's (post-)colonial political history is characterized by interlocking systems of exclusion based on gender, ethnicity, race, class, nationality and locality. These interlocking systems of exclusion that define political citizenship in Ecuador 'are produced and maintained [but also contested and renegotiated] through practices that operate at and across different spatial scales' (McDowell 2008: 496).

Feminist scholars have long critiqued the way political citizenship has been constructed – as in the Ecuadorian case – in masculine and Eurocentric terms (Pateman 1988, Pateman and Mills 2007, Young 1989). In consequence, feminist political geographers have called to broaden the understanding of what gets counted as political space, extending the notion of political participation beyond the sphere of institutionalized politics (Brown and Staeheli 2003, England 2003, Kofman and Peake 1990, Staeheli and Kofman 2004b). While this is a positive move, it has unfortunately resulted in neglecting institutionalized politics in feminist research as a crucial sphere for feminist and gendered politics (for an exception see Secor 2004). This chapter refocuses attention to electoral politics as a site where policies are deliberated that affect women's lives.

This chapter deals with the space of *local* politics in Ecuador. Local politics is not just played out in a metaphorical space when talking about the space women have gained in electoral politics through their increasing presence and their greater share of political power. Rather, the chapter is especially interested how women change the real, material spaces e.g. of the municipality by incorporating a women's office in the building of the municipality. The spaces of politics not only include the buildings of local politics such as the town hall or the provincial council but also other public venues such as plazas, schools, stadia or just the green field that are turned into spaces of politics through the (inter-)action of elected politicians. I focus on local rather than national spaces of politics as in the Ecuadorian context, indigenous people and women have first gained ground in local politics (Arboleda 1993, Brito Merizalde 1997, Radcliffe, Laurie, and Andolina 2002, Van Cott 2008). Hence, the impact of the so-called new political subjects' governance is most visible at the local level where indigenous, Afro-Ecuadorian, and female politicians have controlled some governments since the 1990s. As different regions in Ecuador were affected in different ways by both colonial and postcolonial institutions and as ethnic groups are unevenly dispersed across space due to the country's (post-)colonial history, systems of exclusion interlock in different ways in different political localities. Hence, the particular history of each political locality shapes access to the spaces of local politics along different categories of exclusion and inclusion.

I engage in the following with research that addresses affirmative measures to overcome intersectional exclusions in electoral politics. In general, Holmsten (2010: 1182) points out that 'the vast literature on female and minority representation has largely ignored this question [of intersectionality]'. Institutional approaches that focus on the impact of affirmative action measures such as reserved seats and gender or ethnic quotas have mostly examined female *or* minority representation in isolation from each other. In consequence, a huge body of literature deals exclusively with the impact of gender quotas on women's electoral representation (e.g. Krook 2009, Peschard 2003, Ríos Tobar 2008) and a comparable vast body of literature studies ethnic minority representation (e.g. Rice and Van Cott 2006, Yashar 2006b). Increasingly, however, electoral studies engage with the concept of intersectionality. Grounded in black and multiracial feminist thought (Combahee River Collective 1977, Crenshaw 1989), theories of intersectionality have urged to focus on

> 'the way in which race (or ethnicity) and gender (or other relevant categories) play a role in the shaping of political institutions, political actors, the relationships between institutions and actors, and the relevant categories themselves' (Hancock 2007b: 67).

In the meantime, electoral studies progressively engage with the question how 'sexism, racism and other forms of bigotry [...] create multiple barriers to power' (Hughes 2011: 1). Thereby, research has addressed mainly the electoral representation of minority women, finding that also minority women could theoretically benefit from either gender or minority quotas, in practice often they benefit from neither (Hancock 2007b, Hughes 2011). It has also been shown that minority women's multiple identities can sometimes provide them with opportunities when they strategically emphasize their gender or ethnic minority status in different institutional contexts to enhance electability (Fraga et al. 2008).

The question how different spatial contexts shape processes of identification and dis-identification has also been central in feminist geographies' debates about intersectionality (McDowell 2008, Nightingale 2011, Valentine 2007). Feminist geographers have made fruitful contributions by revealing how particular intersectionalities 'emerge and unfold in different spatial contexts' (Valentine 2007: 15). While most of these studies focus on the intersectionality of identities, geographers haven't engaged yet in debates how context-specific interlocking systems of discrimination can be tackled through political action. Being interested in the intersectionality of social structures rather than identities, I ask how structural intersectionalities can be addressed not through universal recipes but through situated policies 'sensitive to the constitutive roles of spatiotemporal context' (Peck 2011: 773).

Hancock's (2007b) distinction between unitary, multiple and intersectional measures serves as a point of departure for the endeavor to ground policies addressing social inequalities in space and time. A unitary approach implies that one single category is regarded as the most relevant to address. This strategy is often labeled as identity politics. A multiple approach recognizes several categories, but they are treated as conceptually independent. An intersectional approach, in contrast, focuses on the dynamic interaction between different categories, without prioritizing one category as more relevant over others.

Hancock's approach serves in the following first to evaluate the effect of the unitary gender quota and second to discuss the political agendas of local female politicians. By so doing, I am especially interested in the effects of the policies on the very constitution of the spaces of local politics. Thereby, I analyze the impact of policies with regard to the question how through promoting or denying the access of certain gendered, racialized and classed subjects and the political agendas set, the spaces of politics are racialized, gendered and classed in a particular way. Before engaging with the intersectionality of local politics, I introduce briefly a methodological framework to study intersectionality in local politics.

Conceptual differences among approaches to the study of race, gender, class and other categories of difference in political science			
	Unitary Approach	Multiple Approach	Intersectional Approach
Q1: How many categories are addressed?	One	More than one	More than one
Q2: What is the relationship posited between categories?	Category examined is primary	Categories matter equally in a predetermined relationship to each other	Categories matter equally; the relationship between categories is an open empirical question
Q3: How are categories conceptualized?	Static at the individual or institutional level	Static at the individual or institutional level	Dynamic interaction between individual and institutional factors
Q4: What is the presumed makeup of each category?	Uniform	Uniform	Diverse; members often differ in politically significant ways
Q5: What levels of analysis are considered feasible in a single analysis?	Individual *or* institutional	Individual *and* institutional	Individual *integrated* with institutional
Q6: What is the methodological conventional wisdom?	Empirical or Theoretical; Single method preferred; multiple method possible	Empirical or Theoretical; Single method sufficient; multiple method desirable	Empirical and Theoretical; Multiple method necessary and sufficient

Figure 27: Conceptual differences among approaches to the study of race, gender, class and other categories of difference in political science
Source: Hancock 2007b: 64

RESEARCHING INTERSECTIONALITY IN ELECTORAL GEOGRAPHY

'[I]ntersectionality in electoral politics is definitely rich in scholarly promise, if one is willing to take on the challenges of its messy characteristics' (Smooth 2006: 413).

Since intersectionality has been heralded as the 'most important contribution that women's studies has made so far' (McCall 2005: 1771), discussions have emerged about *how* to study intersectionality methodologically. It is the complexity inherent in the concept of intersectionality that turns any empirical attempt to grasp the intersectionality of identities or social structures into quite a 'messy' business (Smooth 2006: 413). When undertaking intersectional research, a whole set of questions raises: Which categories or social structures should be considered? How many of them can be handled in a practical way? Feminist geographers have further asked: Is 'a focus on "sites" (the family, the locality, workplaces)' (McDowell 2008: 502) and the comparison between different sites a way to engage with questions of intersectionality?

So far, intersectional studies in human geography have mostly drawn on qualitative research methods, asking 'what identities are being "done", and when and by whom [...] in specific contexts' (Valentine 2007: 15). Research in feminist geography on intersectionality is predominantly based on interviews with or ethnographic observation of small groups of individuals or communities, conceptualizing intersectionality as lived experience of a particular group (Lundström 2010, Nightingale 2011, Riaño 2011, Valentine 2007). Quantitative studies like

McCall's (2001) study of the intersectional effects of economic restructuring in the US are less common.

Along with Winker and Degele (2011), I advocate the combination of quantitative and qualitative methods to tackle the complexity of intersectional social inequalities. I understand intersectionality as a system of interaction between inequality-creating social structures (colonialism, racism, heteronormativism etc.) and context-specific practices that reproduce or challenge these structures. A multi-method approach is beneficial to grasp both structural inequalities and everyday practices. While the quantitative analysis examines structural discrimination along predefined categories, the interviews with women politicians and ethnographic observation of their political practices focus on categories articulated in women's narrations and addressed through their actions. By linking the lived experiences of the women to structural inequalities, I aim to counterbalance a tendency in poststructuralist research highlighted by Valentine (2007: 19):

> '[I]n our contemporary concern to theorize the intersection of categories we must not lose sight of the fact that the specific social structures of patriarchy, heteronormativity, […] that so preoccupied feminists of the 1970s still matter'.

Winker and Degele (2011: 55) point out that (the interrelatedness of) structural inequalities can be best empirically observed from a historical perspective. Looking back at Ecuador's history of political citizenship, gender, race, class and locality can be identified as central categories that structure the access to electoral participation (Prieto 2004, Prieto and Goetschel 2008). In consequence, the quantitative analysis examines the interrelations between these four categories in local electoral representation. The analysis is based on data from the National Council of Elections (CNE, *Consejo Nacional Electoral*). As data quality provided by the CNE gives only information about the gender of the candidates and not their ethnic or class background, I carried out a representative study about the gender, ethnic, educational and generational composition of local governments in collaboration with the German Development Cooperation and the National Council for Local Governments (GIZ and CONAJUPARE 2009). A questionnaire was sent out to all 798 rural local governments in Ecuador, of which 409 (52 percent) were returned. The data was analyzed against the background of additional statistical data regarding the gender, ethnic and class (levels of poverty and education) composition of the different political localities. This quantitative analysis aims to show the intersectional effect of the unitary gender quota policy on descriptive representation.

On the basis of the quantitative findings, three case study areas were selected. The selection was based on the two following criteria: the presence of female, indigenous and Afro-Ecuadorian local politicians in leading executive positions (basically prefect and mayor) and regional differences that are decisive for the ethnic composition of Ecuador's provinces. Researching the intersectionality of local politics from particular sites made it possible to engage with the way 'intersectionality works out in different places in different ways' (McDowell 2008: 504). To gain insights how the gender quota has (or has not) transformed substantive representation in local politics, extending fieldwork was carried out in each of the three provinces. Between 2006 and 2010, I spent a total of eighteen months in the provinces of Esmeraldas, Chimborazo and Orellana.

In each province, interviews were realized with female politicians elected at the parish, municipal and provincial level. In total, semi-structured, partly biographical interviews were realized with 43 local female politicians and 15 women from local women organizations. Further, I accompanied two prefects, two mayors and four councilors for several weeks in their everyday political activities.

The analysis of the interviews and field notes paid special attention to the subject positions the women politicians referred to in their own narratives and to the social inequalities they address through their policies. Focusing on the way women frame their political agendas on basis of certain traits, the coding of interviews and field notes resulted in a typology of the women's political agendas. The typology is presented through anchor quotes and in form of word clouds. A word cloud is a special visualization of text in which the most frequently used words are effectively highlighted by occupying more prominence in the visual representation. All interview passages in which women were drawing on a specific political issue have been coded and processed in a word cloud generator. While the word clouds show the words women use most frequently when talking about this specific political agenda, each word cloud is labeled according to the trait around which women frame their political agenda.

GENDER QUOTAS AS A MECHANISM FOR ENRICHING DEMOCRATIC INCLUSION?

Quotas are a popular measure to advance the representation of marginalized groups in electoral politics. Gender quotas are the most prevalent. By now, more than 100 countries have adopted gender quotas in some form (Krook 2009). While in general the success of gender quotas is recognized, it has been shown that the quota benefits some women more than others (Hughes 2011). The case of Ecuador serves to critically evaluate the effects of the quota on women's electoral representation. I restrict my analysis to the actual election of candidates and do not take into account other forms of electoral participation such as canvassing or voting.

The *Ley Amparo Laboral* in 1997 and later the National Constitution of 1998 (Art. 102) introduced the gender quota law on the demand of a mainly urban, *mestizo* women's movement (about the introduction of the quota law see Vega Ugalde 2005). Since it was first applied in the elections of 1998, the number of elected female politicians has significantly increased at all political levels (Fig. 3). Analyzing the electoral results since Ecuador's return to democracy in 1979 from a gender perspective, the impact of the quota becomes visible: Whereas women presented around five percent of all councilors in provincial and municipal councils in the first years of democracy and their representation stayed under eight percent during the 1980s and 1990s, their share has considerably risen to 28 percent in 2009 (see Figure 7).

While this story sounds like a success story, a closer look complicates the picture. At the municipal and provincial level, women's participation differs between executive (mayor, prefect) and legislative offices (rural and urban councilor). Women present a significant higher share in legislative councils than in executive posts, being relegated to positions with less power of decision-making and control over budgets. Women's significantly higher presence in municipal and parish councils actually shows the effects of the quota: The quota is only applied to the lists of multi-personal candidacies of legislative bodies and not to unipersonal executive offices such as mayors and prefects. In 2009, only 2 out of 24 prefects (8.7 percent) and 14 out of 221 mayors (6.3 percent) were women. The reorganization of the provincial council in the new Constitution of 2008 further consolidates men's domination. According to the old Constitution, provincial councilors were elected in multi-personal elections in conformation with the gender quota law. Art. 233 of the new Constitution, however, defines that the municipal council appoints half of the provincial councilors. In the most cases, the predominantly male mayors are appointed to the provincial council by the municipal councilors.

Looking at the electoral outcome of 2009 (see Table 8), women's electoral participation is further characterized by an urban/rural cleavage. Political offices representing urban districts such as the urban municipal councilor have a higher percentage of female officeholders than rural ones such as the rural municipal councilors (*consejales rurales*) on cantonal level and representatives of the local rural governments (*Juntas Parroquiales Rurales*) on the level of parishes.

Women in executive and legislative positions			
	Women	Men	Women's share
Prefect (level of province)	2	21	8.7%
Mayor (level of municipality)	14	207	6.3%
Urban municipal councilor	319	721	30.7%
Rural municipal councilor	133	407	24.6%
Local rural governments (parish)	873	3107	21.9%

Table 8: Women in executive and legislative positions in Ecuadorian local politics, 2009-2013
Source: CONAMU 2009

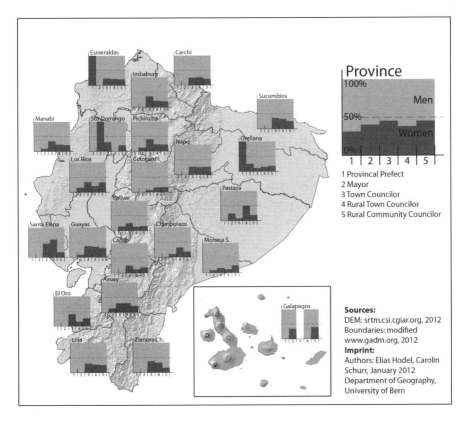

Map 2: Women in executive and legislative positions in Ecuadorian local politics, 2009-2013
Source: Own compilation based on data from CONAMU 2009

Population data indicates that the gap between women's share on urban and rural political offices correlates with the ethnic composition of rural and urban areas. Indigenous people present a higher proportion on the rural population (26 percent) than the urban one (13 percent) (UNDP 2002: 37). When looking at the ethnic identity of politicians, the only data available differentiated in terms of ethnic self-identification is at the level of the rural parishes (*Juntas Parroquiales Rurales, JPR*) (GIZ and CONAJUPARE 2009). The representative census conducted in collaboration with GIZ and CONAJUPARE in the JPRs shows that even though indigenous people present in average 26 percent of the rural population, they present only 13 percent of the rural council members (see Table 8). Hence, it can be said that indigenous people in general are underrepresented in comparison to *mestizo* populations. Afro-Ecuadorian people in contrast present around five percent of the council members and are hence proportionally represented as it is estimated that five percent of Ecuador's total population self-identify as Afro-descendant. Considering that it is to 85 percent men who head the rural local governments (see Table 9), it can be concluded that indigenous women are even further underrepresented in comparison to indigenous men and *mestiza* and Afro-Ecuadorian women. The map of women's electoral participation (see Map 2) shows that women's electoral participation is particularly low in provinces with high incidences of indigenous populations such as the highland provinces of Cotopaxi, Chimborazo and Tungurahua.

	Total in population	Total members JPR	President of the JPR	Vice-President of the JPR	First council member	Second council member	Third council member
Ethnicity							
Indigenous	7 % - 30% (depending on sources)	13,2%	11,3%	12,8%	12,4%	14%	15,7%
Afro-Ecuadorian	7%	4,9%	5,2%	5,4%	4,9%	4,1%	4,6%
Mestiza/o	72%	76 %	76,8%	76,4%	76,2%	76%	74,3%
White	6%	2,3%	3,4%	2,1%	2,6%	1,8%	1,6%
Other	1%	3,7%	3,2%	3,3%	3,9%	4,1%	3,8%
Gender							
Women	50,44%	29,4%	15%	27,7%	35,2%	34,9%	35,2%
Men	49, 56%	70,6%	85%	72,3%	64,8%	65,1%	64,8%

Table 9: Ethnic and gender composition of the Juntas Parroquiales Rurales
Sources: Own compilation based on CEPAL-BID 2009, GIZ und CONAJUPARE 2009, Censo de Población y Vivienda 2009

Table 9 shows that both women and indigenous people occupy rather the less influential posts of the council members and that the position of (vice-)president are still dominated by *mestizo* men. Radcliffe (2010) has shown in her study about the educational biographies of indigenous female leaders that the political underrepresentation of women and especially indigenous women is closely linked to gendered, racialized and urban/rural education opportunities. While illiteracy affects 9% of the total national population, 39 % of indigenous women living in rural communities are illiterate in comparison to only 22% of indigenous men in the same circumstances (CEPAL-BID 2005: 66). Radcliffe's (2010: 327) observation that indigenous women in leadership positions have significant higher levels of education than the average of indigenous women shows the key role education plays in accessing electoral posts. Most of the indigenous women interviewed both in Radcliffe's and my own research have completed secondary education and in some cases even hold university degrees. Poverty is the main reason why girls don't receive education or drop out early. High incidences of poverty in indigenous rural households of 95% (in comparison to a national average of 61%) severely restrict indigenous women's access to education and in consequence to electoral politics. Additionally, indigenous women's electoral participation is further complicated through high levels of institutional violence indigenous women face both in indigenous organizations (Lavinas Picq 2009) and women's organizations (Radcliffe 2002).

The discussion of the differences found in the descriptive representation within local politics has aimed to show how gendered, ethnic, class, rural/urban power relations still saturate the local spaces of politics despite the introduction of a gender quota. The structural intersectionalities that shape the access to electoral politics hit some women (like rural, poor, and indigenous women) harder than others. Therefore, 'compared to women's overall political gains' is 'the presence of indigenous women in public office as a result of popular election disappointingly small' (Pacari 2005: 76). While the electoral representation of women has increased, indigenous women still face a 'patriarchal and mono-ethnic [political] system', as Nina Pacari (2005: 73), the first indigenous female minister in Ecuador's parliament, highlights. It can be concluded that inclusionary policies like the gender quota that do not take into account the intersectionality of exclusions inherent in a polity fail to challenge the structural intersectionalities that shape access to the spaces of politics. The gender quota implemented in Ecuadorian politics has certainly succeeded in increasing women's presence in electoral politics in general. The quality of democracy, however, could further be enriched by fostering the electoral participation of a broader range of women with diverse ethnic, class and local backgrounds that have benefitted to lesser extent from the quota due to structural inequalities.

INTERSECTIONAL POLITICAL AGENDAS?

> 'I do think there is a quantitative effect of the quota, but what are the qualitative achievements of the quota? Although not all women who enter with the quota have a gender agenda, many do advocate a gender agenda and have placed issues like violence against women or social policies that benefit women on the agenda of local politics, issues we never thought could be a topic in local governments' (interview Silvia Vega, 03.08.2010).

Quota policies are generally designed to advance descriptive representation, the 'numeric similarity between legislative bodies and the electorate they represent in terms of gender, race, ethnicity, or other demographic characteristics' (Paxton, Kunovich, and Hughes 2007: 265). In her influential work on 'The Concept of Representation', Hanna Pitkin (1967) has questioned the link between descriptive and substantive representation, arguing that the activities rather than the characteristics of the representatives matter. In a similar vein, Silvia Vega, UNIFEM director and academic scholar, argues in an interview that it is the policies implemented by the women that matter when evaluating the impact of the quota. In this sense, the analysis of political agendas set by female local politicians seems a promising way to study the impact of the quota on substantive representation. The typology of political agendas developed in this section focuses on the question whether the women's agendas address social inequalities through unitary or intersectional policies. I restrict the typology to the three most frequent observed political agendas prioritized by the women on ground of their traits as women, mothers and feminists.

LA MADRE: 'A POLICY FOR THE MOST VULNERABLE'

> 'For the fact that we are mothers, we work for the vulnerable sectors; we like to support those who need our support' (interview Emma Garcia, *mestiza* town councilor Francisco de Orellana, 29.1.2009).

Women with different family status and diverse ethnic and social backgrounds emphasize that their political agendas and practices are closely linked to their identity as women and (potential) mothers. Over half of the 46 interviewed women foregrounded at some point of the interview their motherhood identity as central trait that defines their political action. This observation confirms Craske's (1999: 2) statement that the 'most common identity many female politicians [in Latin America] refer to is that of motherhood'. Pointing out that 'we have this sixth sense, because we have mothers' hearts that God has given to us, we are more reputable, respectable, honest' (interview Prefect Lucia Sosa, 23.4. 2010), many women consider themselves as mothers of their *pueblo*:

> 'My son asked me: "Why is my mum Mayor and my friend's mum has always time for him?" I explained him that I am not only his mum, but also the mother of the whole *pueblo*' (interview Mayor Anita Rivas, 3.2.2009).

When these women talk about their political agendas, they frequently use the word *'pueblo'* (people/town), but also *'familia'* (family), *'hijos'* (children), *'madre'*/*'mama'* (mother), *'corazón'* (heart), *'dios'* (God) or *'amor'* (love) (Fig. 6), responding to the cultural ideal of *marianismo*[17].

---

17 Referring to Catholic imaginations of the holy virgin Maria, *marianismo* was defined by Stevens (1973) as the cult of female spiritual superiority which teaches that women are semi-divine, morally superior to and spiritually stronger than men. Feminist anthropologists have criticized the concept for being an 'ahistorical, essentialist, anachronist, sexist and orientalist fabrication' (Navarro 2002: 270).

Figure 28: Word cloud 'Madre'
Source: Author based on interviews

Like mothers they are concerned about the different – and especially the most vulnerable – members of their 'family', their *pueblo*. They prioritize issues such as education, health, and policies directed at vulnerable sectors that are typically assigned to women within the division of political labor. Mary Mosquera, an Afro-Ecuadorian Vice-Mayor having four children and three grandchildren, is an exemplary case when stating:

> 'I am devoted to social policies; I am also the representative of the cantonal council of children and adolescence. At the moment we try to eradicate children's labor in the rubbish dumps as this is very dangerous for the 75 children who work there' (interview 15.03.2010).

Women are also a central target group of the policies of the women who understand themselves as 'mothers' of the people. *Mestiza* town councilor Lidia Gutierrez in Chimborazo has organized courses for women, in which they learn to sew, knit, and do aerobics. Following the logic of identity politics, they feel as true representatives of the women as the quote from Mary Mosquera highlights:

> 'It is important that there are women in the municipality who represent women, women who have felt the sorrow of the *pueblo*, the sorrow of their *compañera*, who have also suffered under economic constraints' (interview 15.3.2010).

Feeling as true representatives of the women, they are mainly concerned to improve the living conditions of the women in their political localities. In each of the three case study areas, female politicians have initiated trainings for women in hairdressing, beauty, dressmaking, handcrafting or sewing. By so doing, they hope to provide poor women with possibilities to generate or improve their economic income. Balvina Pimbo, the indigenous president of a rural local government in Orellana, for example, has managed to get funding from the oil companies working in her parish for buying sewing machines and training women in sewing the uniforms for the oil companies (interview 26.01.2009). In Esmeraldas and Orellana, local women politicians have further facilitated and financed food markets where women can sell typical food of the region. The female *mestiza* mayor of Francisco de Orellana has opened a small workshop where indigenous women can craft and sell local handicraft to local and (inter-)national tourists at the wharf.

Figure 29: Mayor Anita Rivas in an event with the Association of Kichwa Women
Source: Author

Figure 30: Handicraft center at the wharf in Francisco de Orellana
Source: Author

Women highly appreciate the support of 'their' female mayor as the testimony of the president of the Association of *Kichwa* Women shows:

> 'She [Mayor Anita Rivas] is the only one who has ever supported us and who has given us work, no other mayor has done that. We *Kichwa* women made an effort to organize, but it was the Mayor that helped us' (interview 27.02.2009).

At the first sight, women who build their political action around discourses of motherhood seem to address women's issues – or more general the concern of vulnerable groups – in a unitary way. Women – or disabled or elderly people – are pictured as essentialist, homogenous social groups with specific interests and needs that are addressed by unitary policies like training courses for women. The contents of the registered training courses seem to reproduce a traditional gender order by training women in typical women's issues such as cooking, sewing or handcrafting. This observation suggests that the gender quota essentializes 'woman-ness' as the elected women address primarily women issues and do so based on rather traditional gender roles. The quota has frequently been critiqued precisely on these grounds (Mansbridge 2005).

At the second sight, however, if we focus on who actually benefits from the political activities launched and promoted by the local women politicians, the way many of these policies address structural intersectionalities becomes visible. Activities incentivizing women's economic development address women of social strata that are actually struggling to generate enough income for their family. Hence, these kinds of policies tackle the intersections of class and gender with regard to economic opportunities. Projects like the indigenous handcraft workshop acknowledge that indigenous women suffer particularly under structural discrimination in the labor market due to their ethnic, gender, class, and local position. Hence, in fact many of the activities realized by women understanding motherhood as essential trait that defines their political action can be considered intersectional policies. Each of these political activities is situated in the specific context in which the women act as they respond to necessities that emerge out of the structural intersectionalities that shape these localities. Through these kinds of actions they open up a space for (indigenous) women in public places like the wharf or the central plaza that have post-colonially been reserved mainly for white or *mestizo* men.

LA MAESTRA: 'THE BIGGEST EDUCATIONAL PROJECT IN THE COUNTRY'

'No more teachers without roof, *compañeros*, we are going to dignify, to improve the quality of education. As a teacher, I am going to invest in education' (political speech, Prefect Lucia Sosa, 16.04.2009).

The second frequent trait women base their political agendas on, is the gendered social role of the teacher or educator more broadly. Thereby, education is turned into the center of women's political agendas and the political space constructed by these women is extended towards spaces of education. Especially within local politics, a high percentage of women who run for office still work or have formerly worked as teachers on all educational levels or as social workers more generally. Arboleda (1993: 32) has shown that at the beginning of the 1990s over 60 percent of elected women on the municipal level were teachers. In my own sample of 46 local female politicians, 12 actually have a background in teaching. Many of these women but also women politicians who do not have teaching experience ground their political agenda on the idea that they 'educate' (*educación*) and 'serve' (*servicio, servir*) the 'people' (*el pueblo*) with the same 'responsibility' (*responsabilidad*) as 'teachers' (*maestras*) educate 'children' (*niños*) and 'students' (*estudiantes*).

Figure 31: Word cloud 'Maestra'
Source: Author based on interviews

Political agendas and priorities are defined along their own convictions what the people 'need' to get out of poverty that in their view is education. Prefect Lucia Sosa, who has worked as a teacher in poor rural areas for over 25 years and was also the president of the *Unión Nacional de Educadores* (National Union of teachers), has invested 16 Million US Dollars in the construction of schools in her province. She has carried out 'the biggest educational project in the country's history' (interview Prefect Lucia Sosa, 18.02.2009) thanks to international development funds. Her main aim thereby was that

> ‚the children of the poor rural areas get suitable classrooms (*aulas dignas*) and well trained teachers [...] because it is the least we can do for these children who come from very poor families, in which the parents don't have work' (interview 18.02.2009).

While the magazine of the Provincial Council of Esmeraldas shows her with mainly Afro-Ecuadorian school children (see Figure 32), neither Lucia Sosa nor Vice-Mayor Mary Mosquera discuss issues of education from an ethnic perspective even though census data show that Afro-Ecuadorian people have significantly lower levels of education than *mestizos* (CEPAL-BID 2005: 63) and Afro-Ecuadorian people constitute 44 percent of Esmeraldas' population. They emphasize at various points of the interviews that for them access to education 'is rather a question of class. The majority of poor women and men have the same problems; they are exploited because they don't have education. It is the same class, *negros, cholos, indios, mulatos*, it is one class while the rich are another class' (interview Mary Mosquera, 15.3.2010).

While their focus on class rather than ethnicity is grounded in the socialist rhetoric of their party *Movimiento Popular Democrático (MPD)*, their actions nevertheless do take into account the correlation between poverty and ethnicity, as many of their activities take place in the marginalized *barrios* inhabited mainly by Afro-Ecuadorians.

Figure 32: Prefect Lucia Sosa and her educational project
Source: *Nueva Esmeraldas. Revista anual del Gobierno de la Provincia de Esmeraldas*, Feb. 2010: 35-37

The construction of educational infrastructure also features high in the political agenda of the women politicians in Orellana. The remoteness of many indigenous communities in this Amazon province that are only connected to bigger cities through rivers makes this a challenging and expensive endeavor. Mayor Anita Rivas highlights that the municipality has invested more than two million US dollars in education in 2009, which represents 15 percent of the total municipal budget (interview 17.02.2010). In collaboration with Prefect Guadalupe Llori and with the support of three female town councilors – all of them working as teachers –, she has built class rooms, school canteens, sanitary batteries, playgrounds and houses for the teachers, who often travel long distances to teach the children. Further, Prefect Guadalupe Llori and Mayor Anita Rivas have opened the first university in their province. Town councilor Martha Noboa highlights the importance of this step:

> 'They have opened an extension of the *Escuela Superior Politécnica de Chimborazo* here in Orellana with a department in pedagogy for teachers. This is very important because most of the teachers here have only basic education as the professional ones do not want to work in the remote areas of our province' (interview 12.08.2008).

In Chimborazo, the two indigenous rural town councilors Edith Caranqui and Anna Pilamungo are also worried about the quality of education, especially of bilingual education that teaches children both in Spanish and *Kichwa*, the indigenous language:

> 'Yes, in the [indigenous] *comunidades* we have bilingual schools and this is a big problem. I say that bilingual education has to be of good quality. At the moment we are disadvantaged because the children who graduate in a bilingual school and then go to a *mestizo* secondary school here in the city have lower levels of reading and writing, after six years of bilingual schooling' (interview Edith Caranqui, 14.04.2009).

For the indigenous movement in Ecuador, the implementation of bilingual intercultural education policies was and still is a keystone of indigenous people's new citizenship. The national indigenous organization CONAIE firmed a contract with the National Ministry of Education in 1988 that assigned CONAIE the autonomy to implement and carry out bilingual intercultural education. Martínez Novo (2009b) confirms the problems outlined by Edith Caranqui when pointing out the huge gap between the celebratory discourses of indigenous leaders about the importance of bilingual education to converse the indigenous culture and the poor quality of bilingual education in practice.

As education is often considered a typical women's issue, it is not surprising that education scores high in women politicians' political agendas. The discussed examples of the three case study areas show that female politicians' political actions address different intersectionalities with regard to access and quality of education. The geographical and social context of each province demands different political interventions and the women respond to these demands in different way depending on their ideological, ethnic and class position. Interestingly, gender as a central category that structures access to education was not discussed in any of the three localities. Taking into account that girls' level of education is significantly lower than boys' and contrasting this finding to women's general concern with the situation of girls and women, it is quite surprising that women politicians do not actively address gendered inequalities in education opportunities in their political agendas.

LA FEMINISTA: STRUGGLING FOR GENDER EQUALITY AND AGAINST VIOLENCE'

> 'I am not a feminist in the bad, harmful sense. I am a feminist because I believe in women's value and virtue for social change and there is still a long struggle before us to achieve gender

equality and stop violence' (interview Nina Ruiz, *mestiza* town councilor Riobamba, 04.02.2009).

To fight for the 'rights' (*derechos*) of the 'women' (*mujeres*) for more gender 'equality' (*equidad*) and less 'violence' (*violencia*) (see Figure 33) was the first response of most interviewed women to the question whether they had a particular agenda for women. When talking about their agenda for women, they frequently employed the word 'rights' (*derechos*) as the word cloud shows. By doing so, they emphasized that they ground their political actions that address gender violence and discrimination on a rights-based approach, referring to national and international legislation like CEDAW or the national law against violence (*ley 103*). Political initiatives presented in this section are labeled 'feminist' as they challenge the traditional gender order more explicitly than the former two political agendas as they aim to transform the current patriarchal gender order towards more gender justice rather than reproduce a patriarchal gender order that relegates women to their role as mothers and careers. While many of the women would not position themselves as feminist for the negative stigma or the Western history of the concept, their action can be considered feminist in the way that they fight 'against the institutionalized injustice perpetrated by men as a group against women as a group' – which is an essential characteristic of feminism as defined by Offen (1993: 152).

Figure 33: Word cloud 'Feminista'
Source: Author based on interviews

With 35 of the 46 interviewed women talking about issues of gender violence, this problem certainly is one of women politicians' biggest concerns regarding a feminist political agenda:

'There is lots of exploitation of women in the *mestizo* sector and we are the ones who talk less about it, probably because we feel ashamed, because we don't want that our children know about the violation, I say that out of my own experience, I remained silent, [...]. But this gives me the opportunity to approach the women in the same situation and tell them that we as women, alone or together, have to claim our rights' (interview Lidia Gutiérrez, *mestiza* town councilor Riobamba, 05.08.2008).

Lidia Gutierrez emphasizes that her shared experiences of violence is essential to establish confidence with women who have suffered violence. Hence, rather than the essential trait of being a woman, it is the shared experience that facilitates an emotional bond to female citizens who have suffered violence. This kind of shared experience literally opens up the doors of the town halls to social groups that have not felt welcomed or understood by former male politicians and have therefore not entered the spaces of local politics. It is not only the shared experience, however, that makes women who have suffered violence seek assistance within the spaces of local politics but also actual political actions realized within these spaces. The struggle against violence has long been a concern by Prefect Guadalupe Llori and Mayor Anita Rivas. Due to the men dominated economy of oil extraction that takes place in the Province of Orellana, incidences of sexual violence against women have been particularly high. Guadalupe Llori talks about the long struggle in an interview:

'What did we do? We didn't have a *Comisaria de la Mujer* [legal office where women can report sexual violence] and so we created the *Comisaria de la Mujer*. Later I met the indigenous women group *Alli Guarni Cunas* who asked me for financially support for the women's shelter they wanted to build. Back then I was the Mayor and we build this shelter, because this was a *pueblo machista* who violated the women and even the young girls. I sent 20 workers of the [oil] companies to prison because they had violated an indigenous girl. I fought for that girl until they had all been imprisoned' (interview 22.3.2009).

While *Comisarías de Mujeres* have been built in 21 urban centers right after the 'Law against violence against women and the family' (*Ley 103*) has been passed in the National Congress in 1995, smaller and more rural towns have been slower to transform the law in political reality (Pequeño 2009: 149). Both in Orellana and Chimborazo, local female politicians have played a key role in establishing the *Comisaría de la Mujer* and building women shelters. By so doing, they have changed the vey spatiality of local politics, as in many cases the office of the *comisaría* is hosted in the building of the municipality.

The *comisarías* are open to all women and attend women with very different economic and ethnic backgrounds as violence against women affects women across ethnicity and class. In the already cited interview quote, Lidia Gutierrez problematizes the ethnic dimension of gender violence, suggesting that it is rather recognized as a problem in the indigenous sector and often invisibilized in the *mestizo* sector. In fact, studies show that indigenous women and girls are affected to greater extent from violence than *mestiza* women or girls (Prieto et al. 2010). This finding needs to be contextualized against the backdrop of higher levels of poverty and unemployment that can often be considered reasons for violence against women. Another reason is the patriarchal system of indigenous justice that is still practiced in many indigenous communities. Silvia Vega highlights the importance of strengthening the rights of indigenous women due to this particular system of justice:

> 'The indigenous women have made huge progress in claiming their rights as women within the indigenous culture. Now, our new Constitution respects the system of indigenous justice but emphasizes the rights of indigenous women. The modality of denouncing sexual violence here in Ecuador normally takes place in the *comisarías*, but this does not go with the indigenous culture, because they have the vision to sort this out in the council of elders, the most distinguished members of the *comunidad* call the couple and advise them. But the indigenous women have realized that they as women leave these advisory sessions totally unprotected. But now, they claim rights as indigenous women within the indigenous culture (interview Silvia Vega, 3.8.2010).

Figure 34: Juridical women's office in Riobamba
Source: Picture taken by author

In many *comunidades*, indigenous female politicians have played a major role in organizing women groups in the communities and initiate trainings

> 'where the women learn about their rights, the right of no violence, the right of free maternity treatment, and they receive courses that build up their self-esteem' (interview Anna Pilamungo, 4.4. 2010).

Thanks to the pressure of indigenous women groups, legislation has been forced to take into account the intersectional problems of indigenous women with regard to violence and the indigenous system of justice. Indigenous female leaders and politicians alike now are faced with the challenge to supervise that national legislation is applied in the spaces of local politics and of the *comunidades*.

Figure 35: Indigenous women demanding 'we want a life without violence' at the International Women's Day
Source: Author

As the examples show, local female politicians have played a key role in implementing national laws like the Law against violence against women and the family (*Ley 103*) on local level. The local politicians are challenged to adopt national policies to the specific context of their political territories like the indigenous system of justice or the remoteness of certain communities that make it difficult for women to denounce sexual violence in the urban *comisaria*. By integrating national policies and legislations addressing women's issues and gender equality in local politics, the local female politicians transform and extend local political agendas. In doing so, women politicians expand the boundaries of the spaces of local politics by including issues and social groups that have formerly not been considered and attended within institutionalized local politics. Through their activities that explicitly address women or certain groups of women and often take place in the rooms of the local governments, they gender the spaces of local politics in new ways. Thereby, feminist political agendas challenge the hegemonic gender order more explicitly than policies based on women's trait as mothers or educators as these policies rather reproduce a traditional gender order.

INTERSECTING POLITICAL AGENDAS

'Anita and Guadalupe used to work a lot on issues of gender and have fought for the rights of women. Now with the power, they are worn out – and there are also other issues, they have been working a lot on issues that not only concern women' (interview Juan Antonio Cordoba, 19.01.2009).

The statement from Mayor Anita Rivas' advisor problematizes the typology I have developed in this chapter in two ways: First, it shows that women politicians address a wide range of issues beyond specific women's issues. Second, the political agendas of women are not static, but change over time and across place. The statement highlights that the two women politicians are now rather focusing 'on issues not only concerning women'. Many women politicians emphasize in the interviews that the provision of basic infrastructure such as electricity, water, asphalt, housing etc. are more urgent issues than specific women's issues. Just like

Juan Antonio Cordoba, they do not consider infrastructure projects as women's issues. While frequently only social services such as health and education are framed as women's issues, decisions about infrastructure projects are in fact highly gendered, classed and ethnicized. This becomes clear when analyzing infrastructure project and the investment of local budgets from an intersectional perspective by asking: Who benefits from the construction of a football field, a community center or a bridge? Which *barrio* receives first asphalt or lightning on what grounds? Whose concerns are taken into account in the participatory assembly that decides over budgets?

Women's political agendas constantly change across time and space. Women politicians foreground certain priorities of their political agendas depending on their political (power) position, the current demands of the citizens, the audience they talk to and the places in which they articulate their political program. Accompanying for example Mayor Anita Rivas in her political daily routine, I could witness how she talked about different aspects of her political agenda according to the specific context she was interacting in. She foregrounded her *maestra* identity when opening a refurbished school by positioning herself with many children in front of the cameras, she took on a feminist attitude while discussing with a group of indigenous women what to do against instances of violation of adolescent girls in their *comunidad* when she visited their *communidad*, and she emphasized her mother heart when justifying why she had promised a mother to give her child a scholarship for the local university she had just met on her way to the municipality. Women politicians like Anita Rivas prioritize and articulate different political concerns depending on the specific context they are embedded in and the social group they are interacting with. Depending on the context, they not only emphasize certain initiatives or policies but also play into the foreground certain traits of their own identity such as their socialist political position, their feminist convictions or their motherhood identity. When women politicians prioritize education as most important point on their agenda, they still might support struggles against gender violence or invest in professional trainings for women. In short, their political agendas consist of intersecting and sometimes even paradoxical activities and policies that are shaped by the specific context they are embedded in.

Hence, the question whether and how the women's political agendas bring into being a differently gendered spatiality of politics can only be answered by giving concrete examples that are grounded in specific places. The gender quota law does not transform the spaces of politics per se. Many spaces of politics such as local councils or municipalities are still shaped through political agendas defined by men. If women are elected and present within the local spaces of politics, the way these spaces are (en-)gendered depends on the actual political priorities they advocate.

A GEOGRAPHICAL PERSPECTIVE ON GENDER QUOTAS

This chapter has been concerned with evaluating the impact of the gender quota on local electoral politics in Ecuador. I have argued that quotas and their effects shape and transform the spaces of politics in both their metaphorical and material constitution. Feminist political geographies can fruitfully contribute to questions of electoral representation and participation through their focus on intersectionality, the everyday, the local, and social transformation. By studying the everyday political activities of local women politicians from an intersectional perspective sensitive to the time-spatial context the women are embedded in, a geographical perspective can contribute new insights to the study of the effects of quotas in at least three ways:

First, a geographical approach sharpens the focus on the way spaces of politics are constituted by bodies that are always gendered, racialized, sexualized, and classed in certain ways. Quotas influence the way spaces of politics are gendered, racialized, classed through fostering the presence of bodies that have formerly been excluded from the participation in electoral politics precisely because of the gender or skin color of their body. By focusing on the intersectional identity of the bodies of the elected politicians that constitute the spaces of politics, it can be shown that the quota benefits some women more than others. Focusing on the presence of women in different spaces and scales of politics, the specific time-spatial context of a political locality that shapes access to electoral politics gives evidence why some women benefit less than others from the effect of the quota.

Second, a comparative geographical perspective that evaluates the effect of national policies on local level contributes to develop more effective policies sensitive to the constitutive roles of specific spatiotemporal contexts. The power relations shaping structural intersectionalities of particular localities need to be identified before designing and implementing policies that aim to foster the electoral participation of structurally discriminated social groups. By being sensitive to the structural inequalities that shape local spaces of politics, an intersectional geographic approach can contribute by evaluating which measures work or fail in particular contexts and why.

Third, this chapter has shown that while the unitary approach of the quota fails to tackle the structural intersectionalities that shape access to spaces of electoral politics, women who have been elected thanks to the quota do address structural inequalities through intersectional policies. Geographers are challenged to understand how new political subjects adopt international and national legislations and policies to the specific spatiotemporal context in which they are embedded in. By focusing on the way (inter-)national policies are materialized in the spaces of local politics, it is possible to understand how the gendered and intersectional political agendas of new political subjects transform the very spatiality of local politics. The spatial integration of the *comisarías de la mujer* under the roof of the municipality is an example how the political agendas of new political subjects actually gender the spaces of local politics in new ways.

While the spaces of electoral politics are only one of many spaces where social transformation can be initiated and promoted, they play a crucial role in setting the rules, laws and regulations that organize social life. Even though the quota does not necessarily guarantee a more just, equitable polity in terms of descriptive representation for its unitary focus, the political agendas set by female politicians might nevertheless bring structural inequalities to the center of political attention that affect women in particular ways.

8.
TACKLING FEMINIST POSTCOLONIAL CRITIQUE THROUGH PARTICIPATORY AND INTERSECTIONAL APPROACHES[18]

'Looking back at the first encounters with the local politicians I wanted to make part of my research project, I have to acknowledge that my attempts to act in a decolonizing way totally failed as the following scene "testifies": In Orellana, the female mayor was fascinated from the beginning by my project to analyze how the gender quota law has transformed local politics. She wanted to make me familiar to the rest of the employees in the town hall as soon as possible and thus invited me on my first day to join an internal meeting. She welcomed me and introduced my research, closing with the words "so, we were chosen to be something like her *'conejos de indio'* (indigenous guinea pigs)". My reaction was too slow to respond to this "joke" and I was left behind with the feeling that any attempt to overcome (post-)colonial power relations was a very difficult task' (field notes Schurr, February 2009).

FACING FEMINIST POSTCOLONIAL CRITIQUE

The introductory narrative highlights the dilemma Western researchers face when doing research in the 'Global South'. Since feminist postcolonial scholars have critiqued the colonizing and paternalizing effects of research, Western researchers are constantly challenged to negotiate and reflect on the power relations that constitute the 'field'. The beginning of feminist postcolonial critiques date back to the 1980s, when Chandra Mohanty's 'Under Western Eyes' (1986) and Gayatri Spivak's essay 'Can the Subaltern Speak?' (1988) questioned the authority of Western researchers and their research practices. Mohanty, Spivak and other feminist postcolonial scholars have highlighted the situatedness of knowledge production and critiqued the use of universalizing categories like 'women' or 'people of color'. Feminist postcolonialism engaged in a two-fold project: 'to racialize mainstream feminist theory and to insert feminists concerns into conceptualizations of colonialism and postcolonialism' (Lewis and Mills 2003: 3).

---

18 This chapter has been co-authored with Dörte Segebart and has been published under the same title in the journal Geographica Helvetica (67/3).

While development geography has increasingly incorporated postcolonial thinking (Power, Mohan, and Mercer 2006, Radcliffe 2005, Raghuram and Madge 2006, Sidaway 2007), feminist postcolonial concerns are still relegated to the fringe of the discipline and are mainly addressed by feminist geographers (Laurie and Calla 2004, Radcliffe 2006). Although feminist (postcolonial) debates on issues of fieldwork have been ongoing now for more than two decades (England 1994, Katz 1994, Kobayashi 1994, Nast 1994, Staeheli and Lawson 1994), many of the questions still remain urgent to address: How can research collaboration take place on an equal footing? How can solidarity become the basis of collaborative research? How can we capture the complex entanglements of power relations that saturate people's lives? How can we overcome the burden of postcolonial concerns, which can also block critically engaged scholars, without acting naïvely and being sensitive to existing, new and persisting power relations? In brief: How is it possible to engage with development geography, taking into consideration the (feminist) postcolonial critique?

TOWARDS A FEMINIST POSTCOLONIAL RESEARCH AGENDA

Power asymmetries in knowledge production, especially between Western female researchers and researched non-Western women, have been a central part of feminist postcolonial critiques. In this light, to do research on development issues in the 'global south' as critical feminist researchers, often seems to us to be an impossible endeavor. In this chapter, we would like to reflect on the question that constantly accompanied our research processes in Latin America: How could a feminist *post* postcolonial research agenda look like that facilitates a power-sensitive research process and does not paralyze ourselves as researchers?

What we identify here as a feminist *post* postcolonial research agenda starts with the questioning of simplistic dichotomies such as 'First'/'Third Word' or 'we'/'they' and the call for a more differentiated analysis of power relations. These kinds of essentializing binaries on which early postcolonial thinking was built have been increasingly criticized. Grewal and Kaplan (1994) coin the term 'heterogeneous hegemonies' to highlight the complexity of power relations. Mohanty (2003a, 2003b) also acknowledges the need for more differentiated analysis of power relations in (de-)colonization processes in the revision of her famous paper 'Under Western Eyes' (1986). Ongoing globalization processes make it in Mohanty's (2003b: 506) view necessary to move away from 'geographical and ideological binarisms'. She argues that

> '[w]hile my earlier focus was on the distinction between 'Western' and 'Third World' feminist practices, and while I downplayed the commonalities between these two positions, my focus now is on what I have chosen to call an anti-capitalist transnational feminist practice – and on the possibilities, indeed on the necessities, of cross-national feminist solidarity and organizing against capitalism' (2003b: 509).

Such a 'cross-national feminist solidarity' can be built on ideas of a 'dialogical standpoint theory' as developed by Patricia Hill Collins (1990). She argues that the *one* marginal, critical standpoint that provides the only right perspective on society does not exist and advocates a critical dialogue between positions. This kind of dialogue may lead to the identification of similarities in perspectives that result in politics of solidarity between standpoints and hence to a de-centering of dominant discourses and knowledge claims. Sara Koopman (2011) formulates a similar position as Mohanty, moving towards a political dimension of research. Koopman's call to produce alternative geopolitics in processes of collaborative thinking is in a way refreshing, as it overcomes postcolonial paralysis and instead is nourished by earlier feminist understandings of research as a transformative practice. As she puts it, 'feminist geopolitics is not just about critiquing hegemony, but also about pointing to, and I would agree also *creating*, alternatives' (Koopman 2011: 277). Is she not aware of the inherent power relations in collaborative thinking or in a co-production of knowledge? Or has she already entered a *post* postcolonial phase in which cooperation, solidarity, partnership and common political visions (e.g. peace) seem to be possible? No matter what, the basic question remains, how can we do research in a collaborative and feminist way?

This chapter engages in this debate in order to find constructive ways of conducting feminist postcolonial research. Drawing on our own experiences as German researchers and development practitioners in Latin America, we discuss the potentials and limits of two central feminist postcolonial approaches in development research and practice: participatory (action) research and intersectionality. Our reflections aim to show how development research and practice can benefit from integrating feminist postcolonial approaches.

PARTICIPATORY APPROACHES BETWEEN TYRANNY AND TRANSFORMATION

In this section, we reflect on the potential of participatory approaches, especially of participatory action research (PAR), for feminist postcolonial research. We do so by drawing on Dörte's experience of participatory action research in the context of the participatory monitoring of municipal development plans in the Brazilian Amazon Region. On the basis of this experience, we would like to advocate a differentiated perception of power relations in research, a decolonization of participatory approaches and a systematic integration of PAR in development research as a way to tackle feminist postcolonial critique.

PARTICIPATORY PROCESSES AS RE-COLONIZATION

Participatory approaches reached a relatively prominent position in development practice and research starting in the 1970s (Chambers 1983, Swantz 1970). They seemed to be a solution for overcoming power asymmetries in development practice and research, presenting a potential answer to postcolonial concerns. Power relations, however, evolved and also persisted in these participatory processes (Cooke and Kothari 2001, Guijt and Shah 1998, Nelson and Wright 1995, Pain and Francis 2003).

While in some cases participatory approaches might be empowering to the participants, they are nevertheless increasingly being used as a merely extractive instrument for data collection. Being used in an extractive way, they have rather disempowered than empowered people and created paternalism instead of ownership in many research and development projects. On many occasions, participatory approaches have resulted, instead, in processes of re-colonization: new hierarchies were created and it was mainly the development agencies that benefitted from the time and resources invested by the communities in the hope of realizing the 'perfect' development project – in terms of public image, but also in terms of money and power.

PARTICIPATORY ACTION RESEARCH: RESEARCHING 'WITHOUT GUARANTEES'

As a critical feminist researcher, I (Dörte) wanted to go beyond merely extractivist participatory methods. Participatory action research (PAR) seemed to be a solution, as it encourages the co-production of knowledge by academic researcher and (non-academic) local groups in a collaborative (co-)research process (for an overview see Reason and Bradbury 2001). PAR permits research to lead to action, and can even be seen as action itself. PAR seemed to combine the goals of feminist and postcolonial concerns: to change power relations in the research process and to stimulate transformative action. It offers a radical challenge to *how* data is collected, *what kind* of knowledge results, and *what impacts* these new knowledges have; as well as *who* steers the process and benefits from the research (Kindon et al. 2009: 90). While PAR appears to be an attractive way to respond to feminist (and) postcolonial claims of decolonizing and breaking down hierarchies of knowledge production, it is difficult to translate PAR's premises into research practice, as the example of my research experiences in Brazil highlights. While attempting to follow the ethical concerns of PAR, I was constantly questioning myself: Who is supposed to have the very first research impulse? What if local partners are not willing (or capable) of formulating clear research priorities? What if it is (or seems to be) up to the researcher to translate problems defined by the local population into research priorities? Could such research still be considered a co-production of knowledge?

> 'Even if I wished for, planned and designed it differently, the steering of the [research] processes lay mostly in my own hands. At least it seemed to me that it was me who was constantly giving impulses for planning, action or evaluation phases. Obviously, I integrated indirectly or only weakly enunciated ideas of the participating people and articulated or reformulated them. I assumed the role of a facilitator in this action research, which made me reflect

critically. Hence, I would rather call my procedure a facilitated action research' (Segebart 2007: 144-145, translated by the author).

Participatory methodologies including PAR are not free of power relations, but are constantly challenged by the problem of how to avoid, or how to take into account, power relations within the process. Power relations prevail on various levels: within the group, institution or community, between group and researcher, as well as between men and women or different ethnic groups (Guijt and Shah 1998). Often power relations emerge out of the institutional and financial setting the participatory project is embedded in.

I had to abandon my high expectations of 'pure' action research and accept my role as a professional academic. A long process of reflection started about what Spivak (1990: 9) called 'unlearning one's privilege as loss'. This activity was an inner process of recognizing history and circumstances of learned habits and prejudices, and unlearning dominant systems of knowledge and representation. To have a person with a professional academic background in a group was just a fact and it does not need to be a source of constant paternalism and dominance within the group – at least if the person has done her/his 'homework' (Spivak 1990) and acknowledges that other group members have other capacities and the potential to contribute to the collaborative process (e.g. high acquaintance with the subject, detailed technical and/or traditional knowledge, knowledge of daily routines, important social networks or political contacts or adapted communication skills). It can even be seen as beneficial or adding potential, for the group in their common collaborative research process.

Nevertheless, the advantage of the academic researcher understanding, participating in and directing the research procedure might result in unequal power relations in the research process by assigning him/her the role of being a group leader. There might be at least two possibilities to come to terms with this challenge. One option would be that the researcher could act as a mentor and assist and build the capacities of the research group members in basic research methodology to help them understand, steer and execute the research process together with the researcher. This is highly time-consuming for all involved and has to be fully desired by the co-researchers. Another, less resource intensive, possibility is that the researcher can act as a facilitator (not leader) of the whole research process. S/he might take some decisions by her/himself (e.g. research impulse and applying for funding or the documentation and dissemination of research results) in some moments of the research process. This is what I would call facilitated action research.

A facilitated action research also makes it possible for academics undertaking participatory research to bridge the 'two conflicting social worlds' (Cancian 1993) they constantly confront. On the one hand, academic scholars have to respond to academic standards and understandings of rigor in order to have a successful career (e.g. to comply with strict time frames and to publish in academic journals). On the other hand, communities frequently question the relevance of academic research to meet their needs, demanding other ways of representing research outcomes such as websites, videos, performances, workshops and even real world transformations (Pain and Francis 2003). PAR challenges current academic practices, such as authorship of research, products by which knowledge is shared, funding schemes (e.g. which might not offer funds for the participating community as 'lay co-researcher') and length of research and qualification periods.

After all these critique and limitations, why would one still consider participatory action research as an adequate feminist postcolonial research practice? As we believe in the potential of mutual learning, in the co-production of knowledge, and have experienced those research contexts as transformative spaces, as 'contact zones' (Askins and Pain 2011), we advocate a more rigid and systematic research on power relations in participatory research setting rather than abandoning participatory approaches all together. Such a kind of research should provide a better theorizing and systematizing of participatory approaches (Kapoor 2002). It may also engage with questions of how transformative social relations can be scaled up (Askins and Pain 2011) as well as how favorable research contexts should be shaped, referring to institutional settings, forms of cooperation or co-research, funding and evaluation of academic work.

When implementing participatory approaches, as a possible answer to feminist postcolonial critique, Pain & Francis (2003: 52) suggest that

> 'we should have no illusions that they present straightforward solutions to the ethical dilemmas surrounding research, to the imperative of making research have real impacts, or to the tensions between critical action and academic research'.

We should instead adopt

> '[a] rigorous reflexivity ... [which] requires a level of open-mindedness that accepts that participatory development may inevitably be tyrannical, and a preparedness to abandon it if this is the case' (Cooke and Kothari 2001: 15).

TAKING INTO ACCOUNT INTERSECTIONALITY

When I (Carolin) studied in Quito in 2004 and worked for the German development cooperation (GIZ) with local governments, I became fascinated by the simultaneous empowerment processes of both the indigenous and women's movements. Talking with *mestiza*, indigenous and Afro-Ecuadorian women being elected into local governments, I developed an awareness of the way gender, ethnicity, and class intersected in the spaces of local governments and resulted in complex power relations (Schurr 2009a). Ecuador's contemporary process of political transformation, therefore, seemed an interesting case to empirically engage with the concept of intersectionality.

When I started my PhD research in 2008, intersectionality was the big 'buzzword' (Davis 2008) in the German-speaking feminist academic community. Having emerged out of Black Feminism (Crenshaw 1989, Davis 1981) and Third World Feminism (Anzaldúa 1987, Mohanty 1986), with some time lag, the concept travelled to Europe as a promising concept to come to terms with multiple oppressions (Knapp 2005). Being committed to feminist postcolonialism, I considered the engagement with intersectionality as a possibility to confront feminist postcolonial critique. In this chapter, I reflect on the potentiality of the concept of intersectionality to address the claims of feminist postcolonialism with regard to positionality and development practice.

INTERSECTIONALIZING POSITIONALITY

> 'I was unable to discard or conceal the multiple aspects of my embodied identity that shaped my interactions in the field. Nor was I able to control the ways in which others positioned me' (Sundberg 2005: 19).

Feminist postcolonial critique of the colonizing effects of Western knowledge production (Parpart 1993, Tuhiwai Smith 1999) has provoked vibrant discussions in human geography throughout the last decades. Debates unfolded about how our multiple yet intersecting identities influence the researcher's access to and experience in the field (Datta 2008, Kobayashi 1994), the collection and analysis of data (Hyams 2004, Sidaway 1992), and the writing itself (Palmary 2011, Radcliffe 1994). In fact,

> 'it is now rare to find fieldwork-based feminist research that does not engage to some degree [...] with a reflexive analysis of how the production of ethnographic knowledge is shaped by the shifting contextual, and relational contours of the researcher's social identity' (Nagar 2002: 179).

Emphasizing how 'different parts of embodied social identity were given prominence in different situations' (Sundberg 2005: 19), these accounts seek to shed light on the way the knowledge production is shaped by 'positions of power due to their [the researcher's and the research partners'] position within specific gender, class, or racial hierarchies' (Mullings 1999: 339). Throughout my research, I had similar experiences to the ones described in these accounts regarding the shifting power relations that saturate our research process. Much has been written about the way our research is restricted by these power relations. Little, however, has been reflected on how our intersectional identities make it possible to build up 'situated solidarities' (Nagar and Geiger 2007: 269) during our research process. Reflecting on the relationships with my research partners, I argue along with Mohanty (2003a, 2003b) that feminist postcolonial research needs to shift from its focus on the differences between 'Western' and 'Third World' women (Mohanty 1986) to the possibilities and necessities of 'cross-national feminist solidarity' (Mohanty 2003b: 509). The following field notes serve as an exemplary reflection of the way my own identity intersects with my interests and political positioning, as well as how rapport can be established through intersecting interests and concerns between the research partners and the researcher.

> 'I have already spent a few weeks in the Amazon town of Francisco de Orellana, when I attend, at the side of Balvina Pimbo the president of a local government, a workshop for female local politicians. The discussion turns around two issues: women's underrepresentation within the rural local governments and the women's concern about dropping prices of agricultural

products. Suddenly, Balvina Pimbo addresses me, asking what the situation is like for women in Switzerland. How have they managed to enter politics and what role do they play in Swiss agriculture? So far, I have told her little about my life in Switzerland, as our conversations focus on her everyday challenges as female president. So far my whiteness and academic background rather distance us. The women are clearly surprised when I tell about the late struggle of Swiss women for suffrage that they gained on a national level only in 1971. Having imagined a highly industrialized, grand-scale agriculture (like in the US) they are even more surprised to hear about the struggles of Swiss farmers working in rural villages in the Alps.

Over time, the relationship with Balvina Pimbo has changed: our conversations turn more and more around the question of how women can be mobilized to run as candidates in local politics, how farmers can cooperate or produce organic food in order to develop a sustainable agriculture in the region, or how the violence against women could be addressed in the remote indigenous communities – somehow my whiteness and the class difference between us seems every time less important (field notes Schurr, February 2009).

The field notes show that at the beginning, the relationships in the field were determined by common or different identity markers such as being a woman or having an academic background (Valentine 2002). In the long run, however, shared interests, political aims or ideas of social justice became decisive. The intersectionality of my own positionality and the way my identity and political positions interconnected with those of my research partners enhanced and shaped my encounters in the field. In hindsight, if I reflect on the key people in my three research sites, it is interesting to see that different identities and interests facilitated the establishment of a rapport with each of these individuals: E.g. shared enthusiasm for feminism determined my relationship with Afro-Ecuadorian Mayor Mary Mosquera in Esmeraldas and passionate discussions over strategies to fight against sexual violence enhanced my friendship with *mestiza* provincial deputy Maria Dolores in Chimborazo, who also runs a women's office. In Orellana, my relationship with Mayor Anita Rivas was centered by our shared fight against the extraction of petroleum in the National Park Yasuní. These relationships and the common political aims they were built on shaped the research process in particular ways in each of these three research contexts: They influenced the selection of people I was introduced to and later interviewed, the activities I was actively involved in, my positioning in the field resulting from the way the key person would introduce me and my research, or the topics discussed in interviews and everyday conversations. The concept of intersectionality has enriched my research process by urging me to focus on both the differences between my research partners and myself and the interconnectedness of our experiences, political aims and everyday struggles.

Leaving the individual level aside, and looking in a comparative manner at the situation of women in electoral politics, the similarities of the challenges women politicians face are striking – despite different historical contexts. By highlighting the commonalities in women's struggle for political rights across different national contexts rather than the differences, a transnational feminist solidarity can be built that challenges exclusions inherent in contemporary political systems and advocates for women's citizens rights on all political levels. The awareness of how our political visions intersect with the struggles of our research partners opens up the possibility to turn research into a collaborative decolonizing project built on 'situated solidarity' (Nagar and Geiger 2007: 269).

INTERSECTIONALIZING DEVELOPMENT PRACTICE

> 'An intersectional approach [...] addresses the manner in which racism, patriarchy, class oppression and other discriminatory systems create inequalities that structure the relative positions of women, races, ethnicities, class and the like' (United Nations 2001).

This introductory quote shows that development agencies generally agree on the need for an intersectional analysis of power relations and the importance of recognizing the multiple identities of target groups. While the commitment to an intersectional approach exists on a discursive level, development agencies struggle to integrate intersectional thinking into their everyday practices. Drawing on a collaborative research project between academic researchers, the German Development Cooperation (GIZ) and the National Association of Local Governments (CONAJUPARE), we discuss the challenges of implementing intersectional thinking in development work.

After discussing the importance of an intersectional analysis to understand the challenges of women and indigenous local politicians with the CONAJUPARE during the presentation of my MA thesis results, I was approached by the GIZ regarding the following concern:

> 'The GIZ wants to respond to demands of the CONAJUPARE who is interested in knowing more about the gender, ethnicity and age of the local politicians. The hypothesis of the CONAJUPARE is that women, indigenous and Afro-Ecuadorian people and young people rather participate in local politics than in high politics. CONAJUPARE is interested to identify these leaders with the aim to provide them training based on their specific necessities and with a gender, intercultural, and intergenerational focus' (intern protocol 19.8.2009).

A methodology was collaboratively elaborated. A quantitative survey was sent out to all local rural governments (JPR) with the aim to obtain detailed information about the gender, ethnicity, level of education, and age of all members of each of the 798 local governments in the country. On the basis of a purposely selected sample, 30 interviews were conducted with female, indigenous and young political leaders and analyzed with an intersectional approach (Winker and Degele 2009). Thanks to the study, the CONAJUPARE obtained detailed information about the composition of the JPRs and the underrepresentation of certain groups. This information was employed to promote and support the integration of underrepresented groups in electoral lists through workshops in the 2009 campaign.

At first glance, this project appears to be a best practice example of how researchers and development and political agencies collaboratively integrate an intersectional perspective into their work. Conducting and evaluating this project, however, we also became aware of the limits and challenges of an intersectional approach in development practice.

On an empirical level, our research was constantly confronted with the question, which identity categories structure the access to and experiences in electoral politics. As with most intersectional studies, we focused on gender, ethnicity, class (level of education) and age. We struggled with the vagueness of the concept that does not define 'which categories to use and when to stop' (Davis 2008: 77). In hindsight, having engaged with the reality of local politics in Ecuador through further research, I think that there are other crucial categories that shape the access to and experience in local politics, such as local belonging, sexuality, and marital status. It is certainly true that 'with each new intersection, new connections emerge and previously hidden exclusions come to light' (Davis 2008: 77), which makes intersectionality also such a productive and exciting concept.

On an operational level, an intersectional analysis aiming to contribute to more differentiated development practices and policies is confronted with the methodological challenge of capturing and disentangling complex power relations. This challenge is often accompanied by high operational costs to run such complex studies.

On an institutional level, issues of intersectionality often literally are caught between two stools as gender issues are often covered by different institutions than (ethnic) minority issues (van der Hoogte and Kingma 2004). Cooperation is necessary to effectively develop policies to address questions of intersectionality. Policies that promote one social group in an essentialist way, such as ethnic and gender quotas or training programs exclusively developed for women or indigenous people, need to be rethought from an intersectional perspective to refine policies and projects.

DEVELOPING TRANSNATIONAL FEMINIST RESEARCH PRACTICES

Despite the challenges of doing fieldwork after the postcolonial turn, we would like to conclude by advocating the possibility and necessity of a feminist *post* postcolonial research agenda. As any critical research, feminist postcolonial research should constantly challenge its own assumptions, theories, methodologies and practices in order to address the discussed limitations and problems. We call for a politicization of research and for an engagement in critical globalization research, which should be informed by interregional entanglements and a critical assessment of our own position and role in it. Discussing the impact of feminist postcolonial concerns on both development research and practice, we have aimed to show the mutual benefit of such a dialogue. For one thing, feminist postcolonial approaches can inspire development cooperation in their attempt to decolonize development practices. For another, changes in the design, realization, and evaluation of development projects and everyday routines of development cooperation can reveal the benefits of feminist postcolonial approaches for social, political, and economic transformation. Hence, feminist postcolonial concerns should result in structural transformations in both knowledge production and development cooperation. These structural transformations include: new funding schemes for research and development cooperation, a rethinking of evaluation criteria for both academic success and development progress, obligatory training in feminist postcolonial thought and reflexivity in academia and development practice.

In this chapter we have discussed the limits and potentials of participatory approaches and the concept of intersectionality in a feminist *post* postcolonial research and development agenda. To conclude, we would like to summarize to what extent both approaches can be considered as 'good' feminist (and) postcolonial approaches. We have advocated *participatory approaches* for their (prospective) transformative potential. While the use of participatory approaches in an extractive way can actually disempower the researched and even re-colonize research relations, a power sensitive approach can de-colonize participation in research processes and in development practices. Participatory action research does address some of the shortcomings of conventional participatory approaches, but it is still not free from establishing a space inhabited by complex power relations and faces some methodological challenges. A more modest facilitated participatory action research, which might include elements of mentoring, may effectually translate the idea of co-production of knowledge into practice. In development research, structural changes are needed in favor of participatory approaches, especially PAR, to create more sustainable conditions for those approaches to grow, and for a feminist postcolonial research agenda to be realized. Critical researchers (and development workers) should not stop engaging with participatory approaches, but should do it in a modest and honest, in short decolonizing, fashion.

The reflections on the use of the feminist postcolonial concept of *intersectionality* in development research and practice highlighted the 'open-endedness' of the concept (Davis 2008: 69). First, intersectionality has been employed as a tool to shed light on the way our own (political) positionality interconnects with the intersectional positionality of our research partners, along common interests, political aims and visions of social justice. We have argued that intersecting identity markers such as our gender, ethnicity or nationality are not decisive for establishing rapport in the field but it is rather the way our identities intersect with our interests and political aims that is important. In contrast, the collaborative study presented in the second section is based on a more conventional understanding of the concept of intersectionality, looking at the intersections of gender, ethnicity, class and age in access to, and experience in local electoral politics.

Along with Davis (2008), we would like to argue that it is the ambivalence and the vagueness of the concept of intersectionality that makes it a productive tool to decolonize development research and practice in alignment with feminist postcolonialism. This vagueness opens up possibilities for a creative engagement with the concept in order to identify inclusions and exclusions along intersecting identity categories, and to become aware of differences and commonalities of the intersecting positionalities between the people we encounter in our research and ourselves. Despite all the critiques, intersectionality provides an instance of 'good' feminist theory in the sense that Butler and Scott (1992: xiii) describe when stating 'feminist theory needs to generate analyses, critiques, and political interventions, and open up a political imaginary for feminism'.

CONCLUSIONS

9.
TOWARDS FEMINIST ELECTORAL GEOGRAPHIES

'Make policy, not coffee' (Feminist political button 1970).

'While the participation of women does not guarantee a more just, equitable polity, the marginalization of women is inherently undemocratic' (Staeheli 2004: 352).

(HOW) DO WOMEN DO POLITICS?

Ten Figure 36: Do women do politics?
Source: La Abeja, Periódico del Centro Ecuatoriano para la Promoción y Acción de la Mujer, No. 5, 19.1.1988, Quito, Ecuador, reprinted in: Goetschel, A.M et al. 2007. De Memorias. Imágenes públicas de las mujeres ecuatorianas de comienzos y fines del siglo veinte. Quito: Trama. P. 24.

In 1988, the *Centro Ecuatoriano para la Promoción y Acción de la Mujer* asked 'Do women do politics?' Showing a number of pictures of women realizing diverse activities such as selling food in the market, protesting, and receiving a university degree, the cover page of the magazine questioned a masculinist definition of what politics is and where politics is located. Through rich empirical accounts, feminist (political) geographers have shown how women do politics in public and private loci, as for example, the plaza, the work place, places of worship or at home through such diverse activities. In a similar vein to the magazine cover, they have called for the need to open up of the boundaries of what is understood as 'political' in political geography and geopolitics and urged to re-locate the political both in public and private spaces.

Ten years later, in 1998, with the introduction of the gender quota law, it was no longer a question in Ecuador *whether* women do politics, but *how* they do politics. My research has engaged with the multiple ways in which women do politics within the spaces of local governments in Ecuador. Pleading for the necessity to refocus on the spaces of institutionalized politics in feminist political geography, I have placed the political practices, agendas, and identities of women politicians in Ecuadorian local politics at the center of my research.

PERFORMING POLITICS

In this first part of the conclusion, I summarize the most important empirical findings concerning the performance of women in Ecuadorian local politics. Drawing on the dual meaning of the word 'performance', I consider first the performance of women in electoral politics with regard to how successfully they have been in winning elections and second their performances on the political 'stage' – understood in Butler's sense as enactment of political subjectivity.

How to evaluate the overall performance of women in Ecuadorian local politics? The analysis of electoral results since 1972, when Ecuador reintroduced democracy after years of dictatorship, has shown that women have gained space in electoral politics on all political levels (see Figure 7). The electoral results demonstrate that the gender quota law has had a decisive impact on fostering women's participation in electoral politics (see chapter 7). Analyzing the collected quantitative and qualitative data from an intersectional perspective, however, it became evident that not all women have benefitted from the quota law to the same extent. Indigenous women and women living in rural areas are underrepresented in comparison to *mestizo* and urban women (see Table 9). Interestingly, Afro-Ecuadorian women, a group experiencing similar discrimination as indigenous women, have a higher share in electoral politics than indigenous women. In chapter 7, I have discussed how the access of women to local politics is shaped through context-specific structural intersectionalities: Education opportunities, experience in social and development organizations (NGOs), the individual biography, and embeddedness in local political structures intersect with place-specific gender and racial hierarchical orders, which often have their roots in colonial and post-colonial times. Place-specific gender orders that differ mainly between rural and urban areas and indigenous, *mestizo* and Afro-Ecuadorian sectors are one reason to explain the differences in women's participation in different places and among differently racialized and classed women.

While political parties promote women candidates by placing them on their lists in the mandatory way, only few parties (e.g. the MPD in Esmeraldas) actually offer women training to foster their strengths as political candidates. In general, development organizations have taken up the responsibility to prepare women for their political office. These training programs mostly treat 'women' in an essentialist way as a homogenous group without taking into account the differences women face due to their local, ethnic, and class position (Segebart and Schurr 2010). As affirmative action policies like the quota and institutions such as the National Council for Women (CONAMU) or the National Council for indigenous issues (CODENPE) focus exclusively on one single identity category, women who experience various forms of discrimination often benefit from neither gender nor minority policies. During my research, however, I identified first traces how ethnic minority women organize themselves to promote their specific intersectional interests. Examples from the highland province Chimborazo evidence that indigenous women have started to challenge the structural intersectionalities they face both in (mainstream) women's and in indigenous organizations by creating the network of Kichwa women in Chimborazo.

My research contributes to a feminist electoral geography by placing an intersectional analysis of electoral results and questions of access to electoral politics more generally, at the center of enquiry. An intersectional analysis offers the possibility to examine the power relations that constitute the spaces of politics as well as the effects of affirmative policies that aim to challenge these power relations in a detailed fashion.

How do women perform political identities on political 'stages'? Throughout my research I have asked if the new political subjects in Ecuadorian local politics actually do politics differently from traditional, white, *mestizo*, male politicians. Being convinced at the beginning of my research that women make a difference in politics and do politics differently, I had to learn during my research that the picture is far more complex. Researching women's performances on the diverse 'stages' of local politics and comparing them to those of their male colleagues, I had to acknowledge that there are more similarities than differences, as I had initially believed. In chapter 4, I have analyzed women's identity performances on the electoral stage in Orellana through visual ethnography. Studying the linguistic and embodied performances of women politicians, I have demonstrated that the women politicians constantly negotiate traditional imaginations about how to perform politics and the identity of a politician and, at the same time, distance themselves from and subverted the same discursive imaginations. Chapters 5 and 6 have engaged further with the question whether women construct different spatialities of/for politics through their performances. Focusing on the emotional performances of electoral candidates, I have shown that the new political subjects draw on the same emotional registers of rage and love that dominate the Manichean populist rhetoric of white and *mestizo* populist politicians. While they use similar scripts (like music, dance) and storylines (Manichean struggle between *el pueblo* and elites) to produce the emotional space of campaigning, they slightly alter the emotionality of populist spaces of campaigning. Embedded in the specific time-spatial context of the Province of Orellana in which the observed women and men performed their emotions, their performances assigned a different quality to the emotions of *rabia* and *amor* through drawing on shared experiences made within this specific local context. Their emotional performances differed from those of populist hegemonic subjects in so far as their performances were gendered and racialized in particular ways.

The case studies about women's identity and emotional performances during the electoral campaign 2009 in Orellana (chapters 4, 5 and 6) and Esmeraldas (chapter 6) serve as examples for a more grounded, embodied, and emotional geography of elections. In a similar vein as feminist political geographers have called for an emotional geopolitics (Pain 2009, 2010), I advocate an emotional and embodied geography of elections. Seeking to understand and incorporate emotions in grounded ways is a first step towards a feminist electoral geography interested in the power relations that saturate the spaces of elections. As electoral geographies are embedded in historical, cultural, economic, social, and spatial power relations, we need to understand how these context-specific power-geometries are both (re-)produced and contested through the emotional and embodied practices of campaigning and politics more broadly.

Finally, we have to ask to what extent new political subjects contribute with their political agendas to processes of social change and decolonialization. Feminist electoral geographies need to deal with institutionalized politics beyond the mere study of electoral results in order to evaluate the impact of elections on political agendas and political change. While chapters 4, 5 and 6 have focused on the way new political subjects construct the spaces of politics through their everyday performances, chapters 3 and 7 have looked at the political agendas and policies promoted by new political subjects and asked to what extent these transform the material construction of the spaces of politics. Indigenous political subjects have challenged the political organization and spatial orders of Ecuador's post-colonial democracy through the discursive and material construction of political spaces *and* spaces of politics as pluri-national and intercultural. Intercultural roundtables, participatory budgeting processes, and the appropriation of the material spaces of politics through indigenous symbols and bilingual signage are just a few examples of how indigenous politicians have decolonized Ecuadorian politics. Chapter 7 engages with women politicians' political agendas and their impact on processes of change. The typology of political agendas I have developed has served to evaluate whether women's agendas address social inequalities through unitary or intersectional policies. Women's political agendas are framed by women's positionality as *madres*, *maestras*, and *feministas*. Due to their context-specific positioning, they prioritize certain political agendas such as social policies addressing the most vulnerable sectors, education and (reproductive) health or intra-family violence. Not only have women politicians included issues such as intra-family violence in their political agendas that have not been tackled by male politicians, who located them in the private spaces of the home, but they have also challenged the spatiality of politics by literally making space for such intimate and apparently private issues in the buildings of municipalities or local councils.

Discussing the transformative impact of these policies on traditional gender and political orders, women politicians *do* challenge these orders in different contexts to different extents. By re-placing political agendas and policies at the center of feminist research, I contribute to feminist electoral geographies that are sensitive to social inequalities and interested in collaboratively developing policies promoting social change.

MAKING SPACE FOR FEMINIST ELECTORAL GEOGRAPHIES

The conceptual contribution of this work is a plea for *making space* for feminist issues, theories and methodologies in electoral geography. Departing from the critique of electoral geography as 'atheoretical, quantitative, and positivist' (Cupples 2009: 111, for a similar critique see Agnew 1990, Johnston and Pattie 2004, Warf and Leib 2011), I have introduced concepts and theories from political, social, and feminist theory into electoral geography in order to theorize the constitution of political subjects and spatialities in electoral politics. My empirical study of Ecuadorian local politics has served to show that electoral geography needs to go beyond the mere analysis of electoral results to understand processes of political change. Mouffe's political theory of antagonism/agonism has been employed to analyze the relationship between social movement politics and institutionalized politics that is crucial to understand recent processes of political transformation in Ecuador. Integrating Mouffe's thinking into electoral geography opens up the narrow focus of electoral geography on elections and its respective public spaces. Conceptualizing electoral processes as outcomes of antagonistic processes of negotiation between 'the political' and 'politics', understood in Mouffe's sense,I have focused attention on the power relations between different political collectives of 'us'/'them' that are materialized in both private and public spaces.

To understand how these political collectives that also shape electoral processes are constructed, I have brought Mouffe's notion of antagonism into dialogue with Butler's concept of performativity. The boundaries between different political collectives in Ecuador's colonial and post-colonial spaces of politics are iteratively (re-)produced, stabilized, challenged and shifted through the identity and emotional performances of new political subjects. Focusing on the way the new political subjects' political practices performatively generate political collectives, the performative approach employed in these studies has facilitated opening up the black boxes still present in electoral studies that often treat political subjectivities and collectivities as pre-given entities.

The integration of poststructuralist theories like performativity or intersectionality into electoral studies further serves to challenge the fixed and static use of identity categories such as gender or ethnicity that dominates conventional electoral geography. So far, these identity categories have often been treated as pre-existing and unquestioned entities, used without further questioning as variables in statistical analysis. The poststructuralist concepts of performativity and intersectionality challenge any essentialist and natural-given notion of identity. Employing these two concepts, my empirical research has shown how political subjectivities are produced through multiple identifications, some of which become politically salient for a time in certain contexts. Through Butler's theory of performativity, identities like gender or race that have legitimized social and political hierarchies across time and space can be destabilized and denaturalized. Such destabilization and denaturalization

> 'can be brought about precisely when one demonstrates that these categories do not refer to 'something' which *is* 'natural' but rather *follow from* contingent, historical, malleable *practices* that could have been otherwise, and could be different in future' (Chambers and Carver 2008: 23, emphasis in original text).

While the concept of intersectionality with its focus on intersecting systems of oppression serves to analyze how certain identities are excluded from and marginalized within spaces of politics, a performative approach opens up horizons for political change. Denaturalizing identities like gender and race through showing their performative character enables us to question current power relations that are still based on a colonial political order established through the naturalization of gender and race. All of the theories and concepts discussed in this book work towards and contribute in particular ways to feminism's main aim: social and political change towards a more inclusive democracy. Mouffe's notion of agonism works toward a democratic politics that channels conflicts through agonistic confrontations; Butler hopes to open up the boundaries of what counts as a political subject through her concept of performativity; Ahmed shows the performative power of emotions for binding bodies into political collectives that bring about political change; Black Feminists challenge intersecting systems of oppression through articulating an identity politics at the intersection of different identity categories. In short, integrating feminist theories into political geography in general and electoral geography in particular serves to shift attention towards political change and processes of decolonialization. In the following table, I present a detailed account of the conceptual approaches developed in my research and summarize how it contributes to the development of feminist electoral geographies.

Proposed approach for a feminist electoral geography		
Introducing feminist issues to electoral geography	**Feminist theories of electoral geography**	**Feminist methodologies for a new electoral geography**
understanding (electoral) politics as taking place in both public and private spaces and as constituted through political actions of institutionalized and social movement politics	'the political' and 'politics' as co-constitutive, antagonistic and agonistic (Mouffe 2005a, 2005b, 2008)	long-term ethnographic engagement facilitates tracing political actions in both public and private space and are suited to identify structural and place-specific power relations
researching masculinity, femininity and gendered political subjectivities as constructed in and through institutionalized politics	performativity of gender and political subjectivity (Butler 1990a, 1997, Chambers and Carver 2008)	(visual) ethnography and biographical interviews offer the possibility to capture both linguistic and embodied identity performances
relations between emotions and 'the political'/'politics'	emotional geographies, emotions and social movement, emotions and politics (Bosco 2007, Brown and Pickerill 2009, Butler 2004a, Goodwin, Jasper, and Polletta 2001, Hoggett 2009, Marcus 2002, Pain 2009, Thrift 2004)	(visual) ethnography and (video) interviews to capture linguistic and embodied emotional performances (Chapter 2, Pink 2004, Simpson 2011)
engaging with the intersectionality of structures, identity politics, experiences and policies	intersectionality (Crenshaw 1994b, Frederick 2010, Hancock 2007b, Hughes 2011, Radcliffe 2010, Simien 2007, Smooth 2006, Yuval-Davis 2006)	structural intersectionalities: analysis of electoral results and socio-economic data; intersectional lived experience: interviews & observation; intersectionality of policies: interviews, observation and study of policy documents (Winker and Degele 2011)
relations between emotions and 'the political'/'politics'	emotional geographies, emotions and social movement, emotions and politics (Hoggett 2009, Pain 2009, Thrift 2004)	(visual) ethnography and (video) interviews to capture linguistic and embodied emotional performances (Chapter 2, Pink 2004, Simpson 2011)
the 'local' as key scale and constitutive element of (electoral) politics	feminist geography (Cope 2004, Hyndman 2004, Mountz and Hyndman 2006, Pain 2009)	carry out fieldwork in local spaces of politics in selected case study localities
working towards social change and more inclusive spaces of democracy	feminist political geography/ theory (Dietz 1987, Lovenduski 2005, Nelson 2006, Robinson 2000, Staeheli 2008, 2011, Walsh 2009)	collaborative research that actively constructs alternative and more inclusive spaces of democracy

Table 10: Proposed approach for a feminist electoral geography

To sum up, I would like to sketch the three main points of my call for feminist electoral geographies.

Making space for feminist issues in electoral geography: I call for extending the narrow focus of conventional electoral geographies on electoral results, electoral districting, and party mobilization by understanding electoral politics as closely linked to both 'the political' and 'politics'. Focusing on the relations between social movement activism, elections, and institutionalized politics of elected politicians offers the possibility to embed the electoral geographies within societal power relations and evaluate the impact of elections on fostering social change. By highlighting the importance of bodies, emotions, and everyday practices for the constitution of political subjectivities and spatialities, a feminist electoral geography turns electoral geography into an exciting and colorful endeavor. Producing more vivid accounts of electoral geographies is a first step towards revitalizing electoral geography as a subject worth studying in (feminist) political geography.

Making space for feminist theories in electoral geography: I have argued in my work that integrating feminist (political) theories into electoral geography is a rewarding undertaking. The theoretical lenses of antagonism, performativity, and intersectionality offer possibilities to reconceptualize key concepts of electoral geography like politics, political subjectivities, and spatialities by understanding electoral politics as embodied political performance that produces discursive and material effects. Feminist theories challenge the static, bounded, and essential notions of politics, identity, and space employed in most (positivist) electoral geography. Highlighting the contingency, instability, and fluidity of political subjectivities and spatialities, feminist electoral geographies are well placed to discuss the gendered, racialized, and sexualized geographies of elections.

Making space for feminist methodologies in electoral geography: The shift towards poststructural electoral geographies make space for greater use of qualitative methods that help to uncover new understandings of electoral processes as grounded, contingent, embodied, and deeply intertwined with societal power relations. Feminist methodologies have a long tradition in employing qualitative methodologies that focus on the grounded, embodied and power-saturated practices that constitute social spaces. A poststructuralist electoral geography can benefit from the experiences of feminist (political) geographers by integrating feminist methodologies and politics of fieldwork into its methodological framework. From the stronger focus on qualitative and ethnographic methods to the reflection on positionality and collaborative work, feminist methodologies have much to offer for an electoral geography, seeking to become more sensitive to the power relations that saturate elections, politics, and the research process itself.

REFERENCES

Adelman, Jeremy. 2006. Unfinished states: Historical perspectives on the Andes. In: Drake, Paul W. and Hershberg, Eric (ed.). State and society in conflict: Comparative perspectives on Andean crisis. Pittsburgh: University of Pittsburgh Press, 41–73.

Agnew, John. 1990. From political methodology to geographical social theory? A critical review of electoral geography, 1960-1987. In: Johnston, Ron; Shelley, Fred; and Taylor, Peter (ed.). Developments in electoral geography. London: Croom Helm, 15–21.

———. 1996. Mapping politics: How context counts in electoral geography. In: Political Geography 15(2): 129–146.

Ahmed, Sara. 2004a. The cultural politics of emotion. Edinburgh: Edinburgh University Press.

———. 2004b. Collective feelings. In: Theory, Culture & Society 21(2): 25–42.

Alcoff, Linda. 1991. The problem of speaking for others. In: Cultural Critique 20: 5–32.

———. 2006. Visible identities: Race, gender, and the self. Oxford: Oxford University Press.

Alsono, Ana María. 1992. Gender, power, and historical memory: Discourses of Serrano resistance. In: Butler, Judith and Scott, Joan (ed.). Feminists Theorize the Political. New York: Routledge, 404–425.

Althusser, Louis. 1971. Lenin and philosophy and other essays. London: New Left Books.

Alvarez, Sonia; Dagnino, Evelina; and Escobar, Arturo (ed.). 1998. Cultures of politics/ politics of culture: Re-visioning Latin American social movements. Boulder, Colorado: Westview Press.

Anderson, Kay and Smith, Susan J. 2001. Editorial: Emotional Geographies. In: Transactions of the Institute of British Geographers 26(1): 7–10.

Andolina, Robert. 1999. Colonial legacies and plurinational imaginaries: Indigenous movement politics in Ecuador and Bolivia. Minneapolis: University of Minnesota. Unpublished doctoral dissertation.

———. 2003. The sovereign and its shadow: Constituent Assembly and indigenous movement in Ecuador. In: Journal of Latin American Studies 35: 721-750.

Andolina, Robert; Laurie, Nina; and Radcliffe, Sarah. 2009. Indigenous development in the Andes: Culture, power, and transnationalism. Durham: Duke University Press.

Andolina, Robert; Radcliffe, Sarah; and Laurie, Nina. 2005. Development and culture: Transnational identity making in Bolivia. In: Political Geography 24(6): 678–702.

Ansolabehere, Stephen and Iyengar, Shanto. 1995. Going negative. How attack ads shrink and polarize the electorate. New York: The Free Press.

Anzaldúa, Gloria. 1987. Boderlands. La frontera: The new mestiza. San Francisco: Aunt Lute Books.

Appadurai, Arjun. 2006. Fear of small numbers: An essay on the geography of anger. Durham and London: Duke University Press.

Appiah, Kwame Anthony. 1991. Is the post- in postmodernism the post- in postcolonial? In: Critical Inquiry 17(2): 336–357.

Araújo, Clara and García, Isabel. 2006. Latin America: The experience and the impact of quotas in Latin America. In: Dahlerup, Drude (ed.). Women, quotas and politics. Oxford: Routledge, 83–111.

Arboleda, María. 1993. Mujeres en el poder local. In: Arboldea, María and Rodriguez, Regina (ed.). El espacio posible: Mujercs en el poder local. Santiago: ISIS International, 20–42.

Askins, Kye and Pain, Rachel. 2011. Contact zones: Participation, materiality, and the messiness of interaction. In: Environment and Planning D: Society and Space 29(5): 803–821.

Austin, John L. 1962. How to do things with words. Oxford: Oxford University Press.

Ayala Mora, Enrique. 2008. Resumen de historia del Ecuador. Quito: Corporación Editora Nacional.
Baird, Barbara. 2006. Sexual citizenship in 'the New Tasmania'. In: Political Geography 25(8): 964–987.
Barnett, Clive. 2004. Deconstructing radical democracy: Articulation, representation, and being-with-others. In: Political Geography 23(5): 503–528.
———. 2008. Political affects in public space: Normative blind-spots in non-representational ontologies. In: Transactions of the Institute of British Geographers 33(2): 186–200.
Barnett, Clive and Low, Murray. 2004. Geography and democracy: An introduction. In: Barnett, Clive and Low, Murray (ed.). Spaces of democracy: Geographical perspectives on citizenship, participation and representation. London: Sage, 1–22.
Beck, Scott and Mijeski, Kenneth. 2000. *Indigena* self-identity in Ecuador and the rejection of *Mestizaje*. In: Latin American Research Review 1: 119–137.
———. 2004. Ecuador's Indians in the 1996 and 1998 elections: Assessing Pachakutik's performance. In: Latin Americanist 67(3/4): 46–74.
Becker, Marc. 2003. Race, gender, and protest in Ecuador. In: Peloso, Vincent (ed.). Work, protest, and identity in twentieth-century Latin America. Wilmington: Scholarly Resources, 125–42.
———. 2008a. Indians and leftists in the making of Ecuador's modern indigenous movements. Durham: Duke University Press.
———. 2008b. Pachakutik and indigenous political party politics in Ecuador. In: Stahler-Sholk, Richard; Vanden, Harry; and Kuecker, Glen David (ed.). Latin American social movements in the twenty-first century. Resistance, power and democracy. New York: Rowman & Littlefield Publishers, 165–180.
Becker-Schmidt, Regina. 2007. 'Class', 'gender', 'ethnicity', 'race': Logiken der Differenzsetzung, Verschränkungen von Ungleichheitslagen und gesellschaftliche Strukturierung. In: Klinger, Cornelia; Knapp, Gudrun-Alexi; and Sauer, Birgit (ed.). Achsen der Ungleichheit. Zum Verhältnis von Klasse, Geschlecht und Ethnizität. Frankfurt a. M.: Campus, 56–83.
Bedford, Kate. 2008. Governing intimacy in the World Bank. In: Rai, Shirin M. and Waylen, Giorgina (ed.). Global governance: Feminist perspectives. New York: Palgrave, 84–106.
Bélanger, Paul; Carty, Roland; and Eagles, Munroe. 2003. The geography of Canadian parties' electoral campaigns: Leaders' tours and constituency election results. In: Political Geography 22(4): 439–455.
Bell, David; Binnie, Jon; Cream, Julia; and Valentine, Gill. 1994. All hyped up and no place to go. In: Gender, Place & Culture: A Journal of Feminist Geography 1(1): 31–47.
Bell, David and Valentine, Gill. 1995. Mapping desire: Geographies of sexuality. London: Routledge.
Belli, Simone; Harré, Rom; and Íñiguez, Lupicinio. 2010. What is love? Discourse about emotions in Social Sciences. In: Human Affairs 20(3): 249–270.
Benavides, Hugo O. 2004. Making Ecuadorian histories: Four centuries of defining power. Austin, Texas: University of Texas Press.
Blunt, Alison and Rose, Gillian (ed.). 1994. Writing women and space. Colonial and postcolonial geographies. New York: Guilford Press.
BMZ. 2007. Millenniums-Entwicklungsziele. Ziel 3: Förderung der Gleichstellung der Geschlechter und Stärkung der Rolle der Frauen. Bonn: BMZ.
Boeckler, Marc and Strüver, Anke. 2011. Geographien des Performativen. In: Gebhardt, Hans; Glaser, Rüdiger; Radtke, Ulrich; and Reuber, Paul (ed.). Geographie. Heidelberg: Spektrum Akademischer Verlag, 663–667.
Bondi, Liz. 1990. Feminism, postmodernism, and geography: Space for women? In: Antipode 22(2): 156–167.
———. 2005a. Making connections and thinking through emotions: Between geography and psychotherapy. In: Transactions of the Institute of British Geographers 30(4): 433–448.

Bondi, Liz; Davidson, Joyce; and Smith, Mick. 2005b. Introduction: Geography's 'emotional turn'. In: Davidson, Joyce; Bondi, Liz; and Smith, Mick (ed.). Emotional geographies. Aldershot: Ashgate, 1–18.

Borchorst, Anette and Teigen, Mari 2009. Who is at issue – what is at stake? Intersectionality in Danish and Norwegian gender equality policies. Presented at ECPR Joint Session of Workshops, Lisbon, 14–19 April 2009, Workshop 14, Institutionalizing Intersectionality.

Bosco, Fernando J. 2004. Human rights politics and scaled performances of memory: Conflicts among the *Madres de Plaza de Mayo* in Argentina. In: Social & Cultural Geography 5(3): 381–402.

———. 2006. The *Madres de Plaza de Mayo* and three decades of human rights' activism: Embeddedness, emotions, and social movements. In: Annals of the Association of American Geographers 96(2): 342–365.

———. 2007. Emotions that build networks: Geographies of human rights movements in Argentina and beyond. In: Tijdschrift voor Economische en Sociale Geografie 98(5): 545-563.

Bourdieu, Pierre. 1998. Practical reason: On the theory of action. Stanford, CA.: Stanford University Press.

Bowleg, Lisa. 2008. When Black + lesbian + woman unequals Black lesbian woman: The methodological challenges of qualitative and quantitative intersectionality research. In: Sex Roles 59: 312–325.

Brown, Gavin and Pickerill, Jenny. 2009. Space for emotion in the spaces of activism. In: Emotion, Space and Society 2(1): 24–35.

Brown, Michael. 1997. Replacing citizenship: AIDS activism and radical democracy. New York: Guilford.

Brown, Michael; Knopp, Larry; and Morrill, Richard. 2005. The culture wars and urban electoral politics: Sexuality, race, and class in Tacoma, Washington. In: Political Geography 24(3): 267–291.

Brown, Michael and Staeheli, Lynn. 2003. 'Are we there yet?' Feminist political geographies. In: Gender, Place & Culture: A Journal of Feminist Geography 10(3): 247–255.

Brownill, Sue and Halford, Susan. 1990. Understanding women's involvement in local politics: How useful is a formal/informal dichotomy? In: Political Geography Quarterly 9(4): 396–414.

Büchler, Bettina. 2009. Alltagsräume queerer Migrantinnen in der Schweiz – Ein Plädoyer für eine räumliche Perspektive auf Intersektionalität. In: Binswanger, Christa; Wastl-Walter, Doris; Bridges, Margaret; and Schnegg, Brigitte (ed.). Gender Scripts: Widerspenstige Aneignungen von Geschlechternormen. Frankfurt a.M.: Campus, 41–60.

Burbano de Lara, Felipe. 2004. El impacto de la cuota en los imaginarios masculinos de la política. In: Cañete, María Fernanda (ed.). Reflexiones sobre mujer y política. Memoria del seminario nacional 'Los cambios políticos en el Ecuador: Perspectivas y retos para las mujeres'. Quito: Konrad Adenauer, Unifem, Cedime, Abya-Yala, 89–94.

Bush, Sarah Sunn. 2011. International politics and the spread of quotas for women in legislatures. In: International Organization 65(1): 103–137.

Butler, Judith. 1990a. Gender trouble: Feminism and the subversion of identity. New York: Routledge.

———. 1990b. Performative acts and gender constitution. In: Case, Sue-Ellen (ed.). Performing feminisms: Feminist critical theory and theatre. Baltimore: John Hopkins University Press, 270-283.

———. 1993. Bodies that matter: On the discursive limits of 'sex'. New York: Routledge.

———. 1995. Subjection, resistance, resignification: Between Freud and Foucault. In: Rajchman, John (ed.). The identity in question. New York: Routledge, 229–249.

———. 1997. Excitable speech. A politics of the performative. New York: Routledge.

———. 1999 [1990]. Gender trouble: Feminism and the subversion of identity. New York: Routledge.

———. 2000. Restaging the universal: Hegemony and the limits of formalism. In: Butler, Judith; Laclau, Ernesto; and Zizek, Slavoj (ed.). Contingency, hegemony, universality: Contemporary dialogues on the Left. London: Verso, 11–33.

———. 2004a. Precarious life: The powers of mourning and violence. London: Verso.

———. 2004b. Undoing gender. New York: Routledge.

———. 2005. Gefährdetes Leben. Politische Essays. Frankfurt a.M.: Suhrkamp.

———. 2009. Frames of war: When is life grievable? London: Verso.

Butler, Judith; Laclau, Ernesto; and Zizek, Slavoj (ed.). 2000. Contingency, hegemony, universality: Contemporary dialogues on the Left. London: Verso.

Butler, Judith and Scott, Joan (ed.). 1992. Feminists theorize the political. New York: Routledge.

Byanyima, Winnie. 2007. Women and leadership: The missed Millennium Development Goal. Trinidad and Tobago: Ministry of Community Development, Culture, and Gender Affairs.

Byron, Margaret. 1993. Using audio-visual aids in geography research: Questions of access and responsibility. In: Area 25(4): 379–385.

Campbell, Jan and Harbord, Janet. 1999. Playing it again. In: Theory, Culture & Society 16(2): 229–239.

Cancian, Francesca. 1993. Conflicts between activist research and academic success: Participatory research and alternative strategies. In: The American Sociologist 24(1): 92–106.

Canessa, Andrew. 2005. Natives making nation: Gender, indigeneity and the state in the Andes. Tucson, AZ: University of Arizona Press.

Cañete, María Fernanda. 2000. La crisis ecuatoriana. Sus bloqueos económicos, políticos y sociales. Quito: CEDIME.

———. 2004a. Las vicisitudes de la aplicación de la cuota electoral en los partidos políticos. In: Cañete, María Fernanda (ed.). Reflexiones sobre mujer y política. Memoria del seminario nacional 'Los cambios políticos en el Ecuador: Perspectivas y retos para las mujeres'. Quito: Abya-Yala, 59–70.

——— (ed.). 2004b. Reflexiones sobre mujer y política. Memoria del seminario nacional 'Los cambios políticos en el Ecuador: Perspectivas y retos para las mujeres'. Quito: Abya-Yala.

Carrière, Jean. 2001. Neoliberalism, economic crisis and popular mobilization in Ecuador. In: Demmers, Jolle; Fernández Jilberto, Alex; and Hogenboom, Barbara (ed.). Miraculous metamorphoses: The neoliberalization of Latin American Populism. New York: Zed Books, 132–149.

Castañeda, Jorge. 2006. Latin America's left turn. In: Foreign Affairs 85(3): 28–42.

CEPAL-BID. 2005. Población indígena y afroecuatoriana en Ecuador: Diagnóstico sociodemográfico a partir del censo de 2001. Santiago de Chile: CEPAL-BID.

Chambers, Robert. 1983. Rural development. Putting the last first. Harlow: Longman.

Chambers, Samuel and Carver, Terrell. 2008. Judith Butler and political theory: Troubling politics. New York: Routledge.

Clark, Kim and Becker, Marc. 2007. Indigenous peoples and state formation in modern Ecuador. In: Clark, Kim and Becker, Marc (ed.). Highland Indians and the state in modern Ecuador. Pittsburgh: University of Pittsburgh Press, 1–21.

Clarke, Harold; Sanders, David; Stewart, Marianne; and Whitley, Paul. 2009. Performance politics and the British voter. Cambridge: Cambridge University Press.

Clarke, Simon; Hoggett, Paul; and Thompson, Simon (ed.). 2006. Emotion, politics and society. New York: Palgrave Macmillian.

Cloke, Paul; May, Jon; and Johnsen, Sarah. 2008. Performativity and affect in the homeless city. In: Environment and Planning D: Society and Space 26(2): 241–263.

Collier, John and Collier, Malcom. 1986. Visual anthropology: Photography as a research method. Albuquerque, NM: University of New Mexico Press.

Collins, Jennifer N. 2001. Opening up electoral politics: Political crisis and the rise of Pachakutik. Paper presented at the international meeting of the Latin American Studies Association, Washington, DC.

Collins, Patricia Hill. 1990. Black feminist thought: Knowledge, consciousness and the politics of empowerment. Boston: Unwin Hyman.
———. 1999. Fighting words. Black women and the search for justice. Minneapolis: The University of Minnesota Press.
Colls, Rachel. 2007. Materialising bodily matter: Intra-action and the embodiment of 'fat'. In: Geoforum 38(2): 353–365.
———. 2012. Feminism, bodily difference and non-representational geographies. In: Transactions of the Institute of British Geographers 37(3): 430–445.
Combahee River Collective. 1977. A Black feminist statement. In: Hull, Gloria T. ; Bell Scott, Patricia; and Smith, Barbara (ed.). All the women are white, all the Black are men, but some of us are brave. New York.
Conaghan, Catherine and de la Torre, Carlos. 2008. The permanent campaign of Rafael Correa: Making Ecuador's plebiscitary presidency. In: The International Journal of Press/Politics 13(3): 267–284.
CONAMU. 2002. Las mujeres en el proceso electoral de Mayo 2000. Quito: CONAMU.
Connolly, William E. 2005. The evangelical-capitalist resonance machine. In: Political Theory 33(6): 869–886.
Conway, Janet. 2008. Geographies of transnational feminisms: The politics of place and scale in the World March of Women. In: Social Politics: International Studies in Gender, State and Society 15(2): 207–231.
Cooke, Bill and Kothari, Uma (ed.). 2001. Participation. The new tyranny? London: Zed Books.
Cope, Meghan. 2004. Placing gendered political acts. In: Staeheli, Lynn; Kofman, Eleonore; and Peake, Linda (ed.). Mapping women, making politics. Feminist perspectives on Political Geography. New York: Routledge, 71–86.
Crang, Mike. 2002. Qualitative methods: The new orthodoxy? In: Progress in Human Geography 26(5): 647–655.
Craske, Nikki. 1999. Women and politics in Latin America. New Jersey: Rutgers University Press.
———. 2003. Gender, politics and legislation In: Chant, Sylvia and Craske, Nikki (ed.). Gender in Latin America. New Brunswick: Rutgers University Press, 19–45.
Cream, Julia. 1995. Re-solving riddles: The sexed body. In: Bell, David and Valentine, Gill (ed.). Mapping desire. London: Routledge, 31–40.
Crenshaw, Kimberlé. 1989. Demarginalizing the intersection of race and sex: A Black feminist critique of antidiscrimination doctrine. In: The University of Chicago Legal Forum 1(1): 139–167.
———. 1993. Beyond racism and misogyny. In: Matsuda, Mari; Lawrence, Charles; Delgado, Richard; and Crenshaw, Kimberlé (ed.). Words that wound. Boulder, CO: Westview, 111–132.
———. 1994. Mapping the margins: Intersectionality, identity politics, and violence against women of color. In: Albertson Fineman, Martha and Mykitiuk, Rixanne (ed.). The public nature of private violence. New York: Routledge, 93–118.
———. 1995. Race, reform, and retrenchment: Transformation and legitimation in antidiscrimination law. In: Crenshaw, Kimberlé; Gotanda, Neil; Peller, Garry; and Thomas, Kendall (ed.). Critical race theory. The key writings that formed the movement. New York: The New Press, 103–126.
Crespi, Muriel. 1981. St. John the baptist: The ritual looking glass of hacienda ethnic and power relations. In: Whitten, N. (ed.). Cultural transformations and ethnicity in modern Ecuador. Chicago: University Press of Illinois Press, 477–505.
Cupples, Julie 2009. Rethinking electoral geography: Spaces and practices of democracy in Nicaragua. In: Transactions of the Institute of British Geographers 34(1): 110–124.
Datta, Ayona. 2008. Spatialising performance: Masculinities and femininities in a 'fragmented' field. In: Gender, Place & Culture: A Journal of Feminist Geography 15(2): 189–204.

Davidson, Joyce; Bondi, Liz; and Smith, Mick (ed.). 2005. Emotional geographies. Aldershot: Ashgate.

Davidson, Joyce and Milligan, Christine. 2004. Embodying emotion sensing space: Introducing emotional geographies. In: Social & Cultural Geography 5(4): 523–532.

Davies, Gail and Dwyer, Claire. 2007. Qualitative methods: Are you enchanted or are you alienated? In: Progress in Human Geography 31(2): 257–266.

Davis, Angela. 1981. Women, race and class. New York: Randon House.

Davis, Kathy. 2008. Intersectionality as buzzword: A sociology of science perspective on what makes a feminist theory. In: Feminist Theory 9(1): 67–85.

de la Torre, Carlos. 2000. Populist seduction in Latin America. The Ecuadorian experience. Ohio: Ohio University Press.

———. 2001. Redentores populistas en el neoliberalismo: Nuevos y viejos populismos Latinoamericanos. In: Revista Espanola de Ciencias Políticas 4: 171–196.

———. 2010. Populist seduction in Latin America – Second Edition. Ohio: Ohio University Press.

Degele, Nina and Winker, Gabriele. 2007. Intersektionalität als Mehrebenenanalyse. In: Feministisches Institut Hamburg. Analyse, Positionen & Beratungen: 1–3.

del Campo, Esther. 2005. Women and politics in Latin America: Perspectives and limits of the institutional aspects of women's political representation. In: Social Forces 83(4): 1697–1725.

Dewsbury, John-David. 2000. Performativity and the event: Enacting a philosophy of difference. In: Environment and Planning D: Society and Space 18(4): 473–496.

Dietrich, Anette. 2000. Differenz und Identität im Kontext Postkolonialer Theorie. Eine feministische Betrachtung. Berlin: Logos-Verlag.

Dietz, Mary. 1987. Context is all: Feminism and theories of citizenship. In: Daedalus 116(4): 1–24.

Dirksmeier, Peter. 2009. Performanz, Performativität und Geographie. In: Berichte zur deutschen Landeskunde 83(3): 241–259.

Dirksmeier, Peter and Helbrecht, Ilse. 2008. Time, non-representational theory and the 'performative turn' – towards a new methodology in qualitative social research. In: Forum Qualitative Social Research 9(2): 55.

Dixon, Deborah and Marston, Sallie. 2011. Introduction: Feminist engagements with geopolitics. In: Gender, Place & Culture: A Journal of Feminist Geography 18(4): 445–453.

Dodds, Klaus and Kirby, Philip. 2012. It's not a laughing matter: Critical geopolitics, humour and unlaughter. In: Geopolitics 18(1): 45–59.

Dore, Elizabeth and Molyneux, Maxine. 2000. Hidden histories of gender and the state in Latin America. Durham, NC: Duke University Press.

Dowler, Lorraine. 1998. 'And they think I am just an old lady': Women and war in Belfast, Northern Ireland. In: Gender, Place & Culture: A Journal of Feminist Geography 5(2): 159–176.

Dowler, Lorraine and Sharp, Joanne. 2001. A feminist geopolitics? In: Space & Polity 5(3): 165–176.

Ecuadorenvivo. 2008. Tituaña: 'Rafael Correa es un Waira pamushka en la política del Ecuador'.

El Guindi, Fadwa. 2004. Visual anthropology: Essential method and theory. Plymouth: Altamira Press.

Emmison, Michael and Smith, Philip. 2000. Researching the visual. London: Sage.

England, Kim. 1994. Getting personal: Reflexivity, positionality, and feminist research. In: The Professional Geographer 46(1): 80–89.

———. 2003. Towards a feminist political geography? In: Political Geography 22: 611–616.

Escobar, Arturo and Alvarez, Sonia. 1992. The making of social movements in Latin America: Identity, strategy, and democracy. San Francisco: Westview Press.

Escobar-Lemmon, Maria and Taylor-Robinson, Michelle. 2005. Women ministers in Latin American government: When, where, and why? In: American Journal of Political Science 49(4): 829–844.

Ettlinger, Nancy. 2010. Emotional economic geographies. In: Smith, Susan; Pain, Rachel; Marston, Sallie; and Jones III, John Paul (ed.). The Sage Handbook of Social Geographies. London: Sage, 237–252.

Faria, Caroline. 2013. Staging a new South Sudan in the USA: Men, masculinities and nationalist performance at a diasporic beauty pageant. In: Gender, Place & Culture: A Journal of Feminist Geography 20(1): 87–106.

Fincher, Ruth. 2004. From dualisms to multiplicities: Gendered political practices. In: Staeheli, Lynn; Kofman, Eleonore; and Peake, Linda (ed.). Mapping women, making politics. Feminist perspectives on political geography. New York: Routledge, 49–69.

Flint, Colin and Taylor, Peter. 2007. Political Geography. World-economy, nation-state and locality. Harlow: Pearson Education Limited.

Fluri, Jennifer. 2011a. Armored peacocks and proxy bodies: Gender geopolitics in aid/development spaces of Afghanistan. In: Gender, Place & Culture: A Journal of Feminist Geography 18(4): 519–536.

———. 2011b. Bodies, bombs and barricades: Geographies of conflict and civilian (in)security. In: Transactions of the Institute of British Geographers 36(2): 280–296.

Foucault, Michel 2005. Analytik der Macht. Frankfurt am Main: Suhrkamp.

Fraga, Luis Ricardo; Martinez-Ebers, Valerie ; Lopez, Linda ; and Ramíırez, Ricardo. 2008. Representing gender and ethnicity: Strategic intersectionality. In: Reingold, Beth (ed.). Legislative women: Getting elected, getting ahead. Boulder, CO: Lynne Reiner, 157–174.

Fraser, Nancy. 2005. Mapping the feminist imagination: From redistribution to recognition to representation. In: Constellations 13(3): 295–307.

Fraser, Nancy and Honneth, Axel. 2003. Umverteilung oder Anerkennung? Eine politisch-philosophische Kontroverse. Frankfurt: Suhrkamp.

Frederick, Angela. 2010. Practicing electoral politics in the cracks: Intersectional consciousness in a Latina candidate's city council campaign. In: Gender & Society 24(3): 475–498.

Freidenberg, Flavia. 2007. La tentación populista: Una vía al poder en América Latina. Madrid: Editorial Síntesis.

Garcia-Aracil, Adela and Winter, Carolyn. 2006. Gender and ethnicity differentials in school attainment and labor market earnings in Ecuador. In: World Development 34(2): 289–307.

Garrett, Bradley L. 2011. Videographic geographies: Using digital video for geographic research. In: Progress in Human Geography 35(4): 521–541.

Geertz, Clifford. 1973. The interpretation of cultures. New York: Basic Books.

Gibson-Graham, Julie Katherine. 1994. 'Stuffed if I know!': Reflections on post-modern feminist social research. In: Gender, Place & Culture: A Journal of Feminist Geography 1(2): 205–224.

Goetschel, Ana María; Pequeño, Andrea; Prieto, Mercedes; and Herrera, Gioconda. 2007. De memorias. Imágenes públicas de las mujeres ecuatorianas de comienzos y fines del siglo veinte Quito: Trama.

Gökarıksel, Banu. 2009. Beyond the officially sacred: Religion, secularism, and the body in the production of subjectivity. In: Social & Cultural Geography 10(6): 657–674.

———. 2011. The intimate politics of secularism and the headscarf: The mall, the neighborhood, and the public square in Istanbul. In: Gender, Place & Culture: A Journal of Feminist Geography 19(1): 1–20.

Goldstein, Kenneth M. and Holleque, Matthew. 2010. Getting up off the canvass: Rethinking the study of mobilization. In: Leighley, Jan E. (ed.). The Oxford Handbook of American elections and political behavior. Oxford: Oxford University Press, 577–594.

Goodwin, Jeff; Jasper, James M.; and Polletta, Francesca (ed.). 2001. Passionate politics: Emotions and social movements. Chicago.

Goss, Jasper. 1996. Postcolonialism: Subverting whose empire? In: Third World Quarterly 17: 239–250.

Gould, Deborah. 2004. Passionate political process: Bringing emotions back into the study of social movements. In: Goodwin, Jeff and Jasper, James M. (ed.). Rethinking social movements: Structure, meaning and emotion. Oxford: Rowman & Littlefield Publishers, 155–176.

Gregory, Derek and Pred, Allan (ed.). 2007. Violent geographies: Fear, terror and political violence. London: Routledge.

Gregson, Nicky and Rose, Gillian. 2000. Taking Butler elsewhere: Performativities, spatialities and subjectivities. In: Environment and Planning D: Society and Space 18(4): 433–452.

Grewal, Inderpal and Kaplan, Caren. 1994. Scattered hegemonies: Postmodernity and transnational feminist practice. Minneapolis: University of Minnesota Press.

Gruszczynska, Anna. 2009. 'I was mad about it all, about the ban': Emotional spaces of solidarity in the Poznan March of Equality. In: Emotion, Space and Society 2(1): 44–51.

Guijt, Irene and Shah, Meera Kaul (ed.). 1998. The myth of community. Gender issues in participatory development. London: Practical Action.

Hall, Stuart. 1996a. When was the 'postcolonial'? Thinking at the limit. In: Chambers, Iain and Curti, Lidia (ed.). The post-colonial question. London: Routledge, 242–260.

———. 1996b. Introduction: Who needs 'identity'? In: Hall, Stuart and du Gay, Paul (ed.). Questions of cultural identity. London: Sage, 1–17.

Haller, Aurian. 2003. Art of the demolition derby: Gender, space, and antiproduction. In: Environment and Planning D: Society and Space 21(6): 761–780.

Hampton, Samuel. 2009. Performance and rhetoric: Assessing political speech. In: Affect 1: 3–25.

Hancock, Ange-Marie. 2007a. Intersectionality as a normative and empirical paradigm. In: Politics & Gender 3(2): 248–254.

———. 2007b. When multiplication doesn't equal quick addition: Examining intersectionality as a research paradigm. In: Perspectives on politics 5(1): 63–79.

Haraway, Donna. 1991. Simians, cyborgs, and women: The reinvention of nature. London: Free Association Books.

Harding, Sandra. 1991. Whose science? Whose knowledge? Thinking from women's lives. Ithaca, New York: Cornell University Press.

Harrison, Paul. 2000. Making sense: Embodiment and the sensibilities of the everyday. In: Environment and Planning D: Society and Space 18(4): 497–517.

Hartsock, Nancy. 1983. The feminist standpoint: Developing the ground for a specifically feminist historical materialism. In: Harding, Sandra and Hintikka, Merrill (ed.). Discovering reality: Feminist perspectives on epistemology, metaphysics, methodology, and philosophy of science. Dordrecht, Boston, London: Reidel, 283–310.

Herbert, Steve. 2000. For ethnography. In: Progress in Human Geography 24(4): 550–568.

Herrera, Gioconda. forthcoming. Sujetos y prácticas feministas en el Ecuador: 1980–2005. Quito: FLACSO.

Herrnson, Paul; Lay, Celeste; and Stokes-Brown, Atiya Kai. 2003. Women running 'as women': Candidate gender, campaign issues, and voter-targeting strategies. In: The Journal of Politics 65(1): 244–255.

Herzig, Pascale. 2006. South Asians in Kenya: Gender, generation and changing identities in diaspora. Münster: LIT.

Herzig, Pascale and Richter, Marina. 2004. Von den Achsen der Differenz zu den Differenzräumen: Ein Beitrag zur theoretischen Diskussion in der geografischen Geschlechterforschung. In: Bühler, Elisabeth and Meier Kruker, Verena (ed.). Geschlechterforschung. Neue Impulse für die Geographie. Zürich: Wirtschaftsgeographie und Raumplanung 33, 21–42.

Hillygus, Sunshine D. 2010. Campaign effects on vote choice. In: Leighley, Jan E. (ed.). The Oxford Handbook of American elections and political behavior. Oxford: Oxford University Press, 326–345.

Hoggett, Paul. 2009. Politics, identity, and emotion. Boulder, CO: Paradigm Publishers.

Holmes, Mary. 2004. Feeling beyond rules: politicizing the sociology of emotion and anger in feminist politics. In: European Journal of Social Theory 7(2): 209–228.

Holmsten, Stephanie; Moser, Robert; and Slosar, Mary. 2010. Do ethnic parties exclude women? In: Comparative Political Studies 43(10): 1179–1201.

hooks, bell. 1994. Teaching to transgress. Education as the practice of freedom. London: Routledge.

Hopkins, Peter and Noble, Greg. 2009. Masculinities in place: Situated identities, relations and intersectionality. In: Social & Cultural Geography 10(8): 811–819.

Howard, Rosaleen. 2009. Beyond the lexicon of difference: Discursive performance of identity in the Andes. In: Latin American and Caribbean Ethnic Studies 4(1): 17–46.

———. 2010. Language, signs, and the performance of power: The discursive struggle over decolonization in the Bolivia of Evo Morales. In: Latin American Perspectives 37(3): 176–194.

Hubbard, Philip. 2000. Desire/disgust: Mapping the moral contours of heterosexuality. In: Progress in Human Geography 24(2): 191–217.

Hughes, Melanie. 2011. Intersectionality, quotas, and minority women's political representation worldwide. In: American Political Science Review 2: 1–17.

Hyams, Melissa. 2004. Hearing girls silences: Thoughts on the politics and practices of a feminist method of group discussion. In: Gender, Place & Culture: A Journal of Feminist Geography 11(1): 105–119.

Hyndman, Jennifer. 2001. Towards a feminist geopolitics. In: The Canadian Geographer 45(1): 210–222.

———. 2004. Mind the gap: Bridging feminist and political geography through geopolitics. In: Political Geography 23(3): 307–322.

James, Stanlie. 1997. Women and politics worldwide. In: Signs: Journal of Women in Culture and Society 22(2): 466–469.

Johnston, Ron and Pattie, Charles. 2004. Electoral geography in electoral studies: Putting voters in their place. In: Barnett, Clive and Low, Murray (ed.). Spaces of democracy: Geographical perspectives on citizenship, participation and representation. London: Sage, 45–66.

———. 2006. Putting voters in their place. Geography and elections in Great Britain. Oxford: Oxford University Press.

———. 2008. Representative democracy and electoral geography. In: Agnew, John; Mitchell, Katharyne; and Toal, Gerard (ed.). A Companion to Political Geography. Oxford: Blackwell, 337–355.

———. 2013. Learning electoral geography? Party campaigning, constituency marginality and voting at the 2010 British general election. In: Transactions of the Institute of British Geographers 38(2): 285–298.

Kafka, Franz. 1998 [1925]. Der Prozess. Stuttgart: Reclam.

Kaiser, Robert and Nikiforova, Elena. 2008. The performativity of scale: The social construction of scale effects in Narva, Estonia. In: Environment and Planning D: Society & Space 26(3): 537–562.

Kampwirth, Karen. 2010. Introduction. In: Kampwirth, Karen (ed.). Gender and populism in Latin America: Passionate politics. University Park, PA: The Pennsylvania State University Press, 1–24.

Kapoor, Ilan. 2002. The devil's in the theory: A critical assessment of Robert Chambers' work on participatory development. In: Third World Quaterly 23(1): 101–117.

Katz, Cindi. 1994. Playing the field: Questions of fieldwork in geography. In: The Professional Geographer 46(1): 67–72.

———. 2007. Banal terrorism: Spatial fetishism and everyday insecurity. In: Gregory, Derek and Pred, A. (ed.). Violent geographies: Fear, terror and political violence. London: Routledge, 347–361.

Kemper, Theodore D. 2001. A structural approach to social movement emotions. In: Goodwin, Jeff; Jasper, James M.; and Polletta, Francesca (ed.). Passionate politics: Emotions and social movements. Chicago: University of Chicago Press, 58–73.

Kerner, Ina. 2010. Verhält sich intersektional zu lokal wie postkolonial zu global? Zur Relation von postkolonialen Studien und Intersektionalitätsforschung. In: Reuter, Julia and Villa, Paula Irene (ed.). Postkoloniale Soziologie. Empirische Befunde, theoretische Anschlüsse, politische Intervention. Bielefeld: Transcript, 237–258.

Kindon, Sara. 2003. Participatory video in geographic research: A feminist practice of looking? In: Area 35(2): 142–153.

Kindon, Sara; Pain, Rachel; and Kesby, Mike. 2009. Participatory action research. In: Kitchin, Robert and Thrift, Nigel (ed.). International Encyclopaedia of Human Geography. Oxford: Elsevier, 90–95.

Klinger, Cornelia; Knapp, Gudrun-Alexi; and Sauer, Birgit (ed.). 2007. Achsen der Ungleichheit: Zum Verhältnis von Klasse, Geschlecht und Ethnizität. Frankfurt a. M.: Campus.

Knapp, Gudrun-Axeli. 2005. *Intersectionality* – ein neues Paradigma feministischer Theorie? Zur transatlantischen Reise von *Race, Class, Gender*. In: Feministische Studien 23(1): 68–81.

Knopp, Larry and Brown, Michael. 2003. Queer diffusions. In: Environment and Planning D: Society and Space 21(4): 409–424.

Kobayashi, Audrey. 1994. Coloring the field: Gender, 'race', and the politics of fieldwork. In: The Professional Geographer 46(1): 73–90.

Kobayashi, Audrey and Peake, Linda. 1994. Unnatural discourse: 'Race' and gender in geography. In: Gender, Place & Culture: A Journal of Feminist Geography 1: 225–243.

Kofman, Elenore. 2008. Feminist transformations of political geography. In: Cox, Kevin; Low, Murray; and Robinson, Jennifer (ed.). The Sage Handbook of Political Geography. London: Sage, 73–86.

Kofman, Elenore and Peake, Linda. 1990. Into the 1990s: A gendered agenda for a political geography. In: Political Geography 9(4): 313–336.

Kondo, Dorinne. 1990. Crafting selves: Power, gender, and discourses of identity in a Japanese workplace. Chicago: University of Chicago Press.

Koopman, Sara. 2011. Alter-geopolitics: Other securities are happening. In: Geoforum 42(3): 274–284.

Korf, Benedikt. forthcoming. Geographie des Denkens. In: Geographische Zeitschrift.

Krook, Mona Lena. 2009. Quotas for women in politics: Gender and candidate selection reform worldwide. New York: Oxford University Press.

Kuus, Merje. 2007. Ubiquitous identities and elusive subjects: Puzzles from Central Europe. In: Transactions of the Institute of British Geographers 32(1): 90–101.

———. 2012. Foreign policy and ethnography: A sceptical intervention. In: Geopolitics 18(1): 115–131.

Laclau, Ernesto. 1990. New reflections on the revolution of our time. London: Verso.

———. 1996. Emancipation(s). London: Verso.

———. 2005. On populist reason. London: Verso.

Laclau, Ernesto and Mouffe, Chantal. 1985. Hegemony and socialist strategy. Towards a radical democratic politics. London: Verso.

Lalander, Rickard. 2010. Between interculturalism and ethnocentrism: Local government and the indigenous movement in Otavalo-Ecuador. In: Bulletin of Latin American Research 29(4): 505–521.

Latham, Alan. 2003. Research, performance, and doing human geography: Some reflections on the diary-photograph, diary-interview method. In: Environment and Planning A 35(11): 1993–2017.

Latham, Alan and McCormack, Derek P. 2007. Digital photography and web-based assignments in an urban field course: Snapshots from Berlin. In: Journal of Geography in Higher Education 31(2): 241–256.

Laurie, Nina; Andolina, Robert; and Radcliffe, Sarah. 2003. Indigenous professionalization. Transnational social reproduction in the Andes. In: Antipode 35(3): 463–491.
———. 2005. Ethnodevelopment: Social movements, creating experts and professionalising indigenous knowledge in Ecuador. In: Antipode 37(3): 470–496.
Laurie, Nina and Calla, Pamela. 2004. Development, postcolonialism and feminist political geography. In: Staeheli, Lynn; Kofman, Eleonore; and Peake, Linda (ed.). Mapping women, making politics. Feminist perspectives on political geography. London: Routledge, 99–112.
Lavinas Picq, Manuela. 2009. La violencia como factor de exclusión política: Mujeres indígenas en Chimborazo. In: Pequeño, Andrea (ed.). Participación y políticas de mujeres indígenas en América Latina. Quito: FLACSO, 125–143.
Law, John and Urry, John. 2004. Enacting the social. In: Economy and Society 33(3): 390–410.
Leib, Jonathan and Quinton, Nicholas. 2011. On the shores of the 'moribund backwater'? Trends in electoral geography research since 1990. In: Warf, Barney and Leib, Jonathan (ed.). Revitalizing electoral geography. Farnham: Ashgate, 9–30.
Lewis, Reina and Mills, Sara (ed.). 2003. Feminist postcolonial theory. A reader. New York: Routledge.
Lind, Amy. 2000. Negotiating boundaries: Women's organizations and the politics of restructuring in Ecuador. In: Marchand, Marianne and Runyan, Anne (ed.). Gender and global restructuring. Sightings, sites and resistances. London: Routledge, 161–175.
———. 2001. Making feminist sense of neoliberalism: The institutionalization of women's struggles for survival in Ecuador and Bolivia. In: Menjívar, Cecilia (ed.). Through the eyes of women: Gender, social networks, family, and structural change in Latin America and the Carribean. Willowdale: de Sitter, 231–261.
Lluco, Miguel. 2005. Acera del movimiento de Unidad Plurinacional. In: Escárzaga, Fabiola and Gutiérrez, Raquel (ed.). Movimiento indígena en América Latina: Resistencia y proyecto alternativa. Puebla, Mexico: Juan Pablos editores, 119–132.
Lobo, Michele. 2010. Negotiating emotions, rethinking otherness in suburban Melbourne. In: Gender, Place & Culture: A Journal of Feminist Geography 17(1): 99–114.
Longhurst, Robyn. 1997. Disembodied geographies. In: Progress in Human Geography 21: 486–501.
———. 2000. 'Corporeographies' of pregnancy: 'Bikini babes'. In: Environment and Planning D: Society & Space 18(4): 453–472.
———. 2001. Bodies: Exploring fluid boundaries. London: Routledge.
Lorimer, Jamie. 2010. Moving image methodologies for more-than-human geographies. In: Cultural Geographies 17(2): 237–258.
Lovenduski, Joni. 2005. Feminizing politics. Cambridge: Polity Press.
———. 2010. The dynamics of gender and party. In: Krook, Mona Lena and Childs, Sarah (ed.). Women, gender, and politics: A reader. Oxford: Oxford University Press, 81–86.
Lucas, Kintto. 2000. We will not dance on our grandparents' tombs: Indigenous uprisings in Ecuador. London: Institute for International Relations.
Lucero, José Antonio. 2008. Struggles of voices: The politics of indigenous representation in the Andes. Pittsburgh: University of Pittsburg.
Lundström, Catrin. 2010. 'Concrete bodies' young Latina women transgressing the boundaries of race and class in white inner-city Stockholm. In: Gender, Place & Culture: A Journal of Feminist Geography 17(2): 151–167.
Lüneborg, Margreth. 2009. Politik auf dem Boulevard? Die Neuordnung der Geschlechter in der Politik der Mediengesellschaft. Bielefeld: Transcript.
Lyons, Barry. 2002. 'To act like a man'. Masculinity, resistance and authority in the Ecuadorian Andes. In: Montoya, Rosario; Frazier, Lessie Jo; and Hurtig, Janise (ed.). Gender's place. Feminist anthropologies of Latin America. New York: Palgrave, 45–64.
Macas, Luis. 2002. Lucha del movimiento indígena en el Ecuador. Quito: Boletín ICCI.

Mahtani, Minelle. 2002. Tricking the border guards: Performing race. In: Environment and Planning D: Society and Space 20(4): 425–440.

Malbon, Ben. 1999. Clubbing. Dancing, ecstasy and vitality. London: Routledge.

Mansbridge, Jane. 2005. Quota problems: Combating the dangers of essentialism. In: Politics & Gender 1(4): 622–638.

Marcus, George E. 2002. The sentimental citizen: Emotion in democratic politics. University Park, PA: Pennsylvania State University Press.

Marcus, George E.; MacKuen, Michael; Wolak, Jennifer; and Keele, Luke. 2006. The measure and mismeasure of emotion. In: Redlawsk (ed.). Feeling politics: Emotion in political information processing. New York: Palgrave Macmillan, 31–46.

Marcus, George E. and Mackuen, Michael B. 1993. Anxiety, enthusiasm, and the vote: The emotional underpinnings of learning and involvement during presidential campaigns. In: The American Political Science Review 87(3): 672–685.

Marcus, George E.; Neumann, Russell W.; and MacKuen, Michael. 2000. Affective intelligence and political judgment. Chicago: University of Chicago Press.

Marschall, Melissa. 2010. The study of local elections in American politics. In: Leighley, Jan E. (ed.). The Oxford Handbook of American elections and political behavior. Oxford: Oxford University Press, 471–492.

Marston, Sallie. 1990. Who are 'the people'? Gender, citizenship, and the making of the American nation. In: Environment and Planning D: Society and Space 8(4): 449–458.

Martínez Novo, Carmen (ed.). 2009a. Repensando los movimientos indígenas. Quito: FLACSO.

———. 2009b. La crisis del proyecto cultural del movimiento indígena. In: Martínez Novo, Carmen (ed.). Repensando los movimientos indígenas. Quito: FLACSO, 173–198.

Massey, Doreen. 1993. Power-geometry and a progressive sense of place. In: Bird, John; Curtis, Barry; Putnam, Tim; Robertson, George; and Tickner, Lisa (ed.). Mapping the futures: Local cultures, global change. London: Routledge, 59–69.

———. 1999. Imagining globalisation: Power-geometries of time-space. In: Massey, Doreen (ed.). Power-geometries and the politics of space-time. Heidelberg: Universität Heidelberg, 9–23.

Mattissek, Annika. 2008. Die neoliberale Stadt: Diskursive Repräsentationen im Stadtmarketing deutscher Grossstädte. Bielefeld: Transcript.

Mayer, Tamar. 2004. Embodied nationalisms. In: Kofman, Elenore; Peake, Linda; and Staeheli, Lynn (ed.). Mapping women, making politics. Feminist perspectives on Political Geography. New York: Routledge, 153–168.

McCall, Leslie. 2001. Complex inequality: Gender, class and race in the new economy. New York: Routledge.

———. 2005. The complexity of intersectionality. In: Journal of Women in Culture and Society 30(31): 1771–1800.

McClintock, Anne. 1995. Imperial leather. Race, gender and sexuality in the colonial contest. New York: Routledge.

McCormack, Derek. 2003. An event of geographical ethics in spaces of affect. In: Transactions of the Institute of British Geographers 28(4): 488–507.

———. 2005. Diagramming practice and performance. In: Environment and Planning D: Society and Space 23(1): 119–147.

———. 2006. For the love of pipes and cables: A response to Deborah Thien. In: Area 38(3): 330–332.

———. 2009. Performativity. In: Rob, Kitchin and Nigel, Thrift (ed.). International Encyclopedia of Human Geography. Oxford: Elsevier, 133–136.

McDowell, Linda. 1992. Doing gender: Feminism, feminists and research methods in Human Geography. In: Transactions of the Institute of British Geographers 17(4): 399–416.

———. 2008. Thinking through work: Complex inequalities, constructions of difference and trans-national migrants. In: Progress in Human Geography 32(4): 491–507.

———. 2009. Working bodies: Interactive service employment and workplace identities. Oxford: Wiley-Blackwell.

McEwan, Cheryl. 2000. Engendering citizenship: Gendered spaces of democracy in South Africa. In: Political Geography 19(5): 627–651.

———. 2005. New spaces of citizenship? Rethinking gendered participation and empowerment in South Africa. In: Political Geography 24(8): 969–991.

Mead, Margarete. 2003. Visual anthropology in a discipline of words. In: Hockings, Paul (ed.). Principles of visual anthropology. Berlin, New York: Mouton de Gruyter, 3–10.

Megoran, Nick. 2005. The critical geopolitics of danger in Uzbekistan and Kyrgyzstan. In: Environment and Planning D: Society and Space 23(4): 555–580.

———. 2006. For ethnography in political geography: Experiencing and re-imagining Ferghana Valley boundary closures. In: Political Geography 25(6): 622–640.

Merizalde, Brito and Soledad, Monica. 1997. La participación de la mujer en la política Ecuatoriana. Quito: Editorial Universitaria Quito.

Mignolo, Walter D. 2005. The idea of Latin America. Oxford: Blackwell.

Mills, Charles. 1997. The racial contract. Ithaca: Cornell University Press.

Mills, Sara. 1996. Gender and colonial space. In: Gender, Place & Culture: A Journal of Feminist Geography 3(2): 125–147.

Mohanty, Chandra. 1986. Under Western eyes: Feminist scholarship and colonial discourses. In: Boundary 12(3): 333–358.

———. 2003a. Feminism without borders. Decolonizing theory, practicing solidarity. Durham: Duke University Press

———. 2003b. 'Under Western eyes' revisited: Feminist solidarity through anticapitalist struggles. In: Signs: Journal of Women in Culture and Society 28(2): 499–535.

Molyneux, Maxine. 2000. Twentieth-century state formations in Latin America. In: Dore, Elizabeth and Molyneux, Maxine (ed.). Hidden histories of gender and the state in Latin America. Durham: Duke University Press, 33–81.

Monk, Janice; Manning, Patricia; and Denman, Catalina. 2003. Working together: Feminist perspectives on collaborative research and action. In: ACME: International Journal of Critical Geography 2: 91–106.

Moraña, Mabel; Dussel, Enrique; and Jáuregui, Carlos. 2008. Colonialism and its replicants. In: Moraña, Mabel; Dussel, Enrique; and Jáuregui, Carlos (ed.). Coloniality at large: Latin America and the postcolonial debate. Durham, NC: Duke University Press, 1–22.

Morrill, Richard. 1981. Political districting and geographic theory. Washington, DC: Association of American Geographers.

———. 1987. Redistricting, region and representation. In: Political Geography Quarterly 6: 241–260.

Morton, Frances. 2005. Performing ethnography: Irish traditional music sessions and new methodological spaces. In: Social & Cultural Geography 6(5): 661–676.

Mosquera Andrade, Violeta. 2006. Mujeres congresistas. Estereotipos sexistas e identidades estratégicas, Ecuador 2003–2005. Quito: FLACSO.

Mosquera Andrade, Violeta; Schurr, Carolin; and CONAJUPARE. 2009. Liderazgo de mujeres, jóvenes y étnicos a nivel Parroquial. Quito: GTZ.

Moss, Pamela. 2003. Feminist geography in practice: Research and methods. Oxford: Blackwell.

Mouffe, Chantal. 1979. Gramsci and marxist theory. London: Routledge.

———. 1993. The return to the political. London: Verso.

———. 1995. Post-Marxism: Democracy and identity. In: Environment and Planning D: Society and Space 13(3): 259–265.

———. 2005a. The democratic paradox. London: Verso.

———. 2005b. On the political. New York: Routledge.

———. 2007. Über das Politische. Wider die kosmopolitische Illusion. Frankfurt am Main: Suhrkamp.

———. 2008. Agonistische Politik, Pluralistische Demokratie und Feminismus. In: Krondorfer, Birge; Wischer, Miriam; and Strutzmann, Andrea (ed.). Frauen und Politik: Nachrichten aus Demokratien. Wien: Promedia Verlag, 35–45.

Mountz, Alison and Hyndman, Jennifer. 2006. Feminist approaches to the global intimate. In: Women's studies quarterly 34(1&2): 446–463.

Müller, Martin. 2007. What's in a word? Problematising translation between languages. In: Area 39(2): 206–213.

———. 2008. Reconsidering the concept of discourse for the field of critical geopolitics: Towards discourse as language and practice. In: Political Geography 27(3): 322–338.

———. 2009. Making great power identities in Russia: An ethnographic discourse analysis of education at a Russian elite university. Zürich: LIT.

———. 2012. Mittendrin statt nur dabei: Ethnographie als Methodologie in der Humangeographie. In: Geographica Helvetica 67(4): 179–184.

Mullings, Beverley. 1999. Insider or outsider, both or neither: Some dilemmas of interviewing in a cross-cultural setting. In: Geoforum 30(4): 337–350.

Nagar, Richa. 2002. Footloose researchers, 'traveling' theories, and the politics of transnational feminist praxis. In: Gender, Place & Culture: A Journal of Feminist Geography 9(2): 179–186.

———. 2013. Storytelling and co-authorship in feminist alliance work: reflections from a journey. In: Gender, Place & Culture: A Journal of Feminist Geography 20(1): 1–18.

Nagar, Richa and Geiger, Susan. 2007. Reflexivity and positionality in feminist fieldwork revisited. In: Tickell, Adam; Sheppard, Eric; Peck, Jamie; and Barnes, Trevor (ed.). Politics and practice in Economic Geography. London: Sage, 267–278.

Nash, Catherine. 2000. Performativity in practice: Some recent work in cultural geography. In: Progress in Human Geography 24(4): 653–664.

Nast, Heidi J. 1994. Women in the field: Critical feminist methodologies and theoretical perspectives. In: The Professional Geographer 46(1): 54–66.

Natter, Wolfgang. 1995. Radical democracy: Hegemony, reason, time and space. In: Environment and Planning D: Society and Space 13(3): 267–274.

Navarro, Marysa. 2002. Against Marianismo In: Montoya, Rosario; Frazier, Lessie Jo; and Hurtig, Janise (ed.). Gender's place. Feminist anthropologies of Latin America. New York: Palgrave, 257–272.

Nelson, Diane M. 1999. A finger in the wound. Body politics in Quincentennial Guatemala. Berkeley: University of California Press.

Nelson, Lise. 1999. Bodies (and spaces) do matter: The limits of performativity. In: Gender, Place & Culture: A Journal of Feminist Geography 6(4): 331–353.

———. 2006. Geographies of State power, protest, and women's political identity formation in Michoacán, Mexico. In: Annals of the Association of American Geographers 96(2): 366–389.

Nelson, Nici and Wright, Susan (ed.). 1995. Power and participatory development. Theory and practice. London: Practical Action.

Neumann, Russell W.; Marcus, George E.; Crigler, Ann N.; and MacKuen, Michael (ed.). 2007. The affect effect: Dynamics of emotion in political thinking and behavior. Chicago: University of Chicago Press.

Ngai, Sianne. 2005. Ugly feelings. Cambridge, MA: Harvard University Press.

Nicley, Erinn. 2011. Elections and cultural political economy: The political geography of the Bloque Nacionalista Galego in the Galicia autonomous community. In: Warf, Barney and Leib, Jonathan (ed.). Revitalizing electoral geography. Farnham: Ashgate, 75–96.

Nightingale, Andrea J. 2011. Bounding difference: Intersectionality and the material production of gender, caste, class and environment in Nepal. In: Geoforum 42(2): 153–162.

Noble, Greg. 2009. 'Countless acts of recognition': Young men, ethnicity and the messiness of identities in everyday life. In: Social & Cultural Geography 10(8): 875–891.

Norris, Pippa and Inglehart, Ronald. 2001. Cultural obstacles to equal representation. In: Journal of Democracy 12(3): 126–140.
O'Connor, Erin. 2003. Indian and national salvation: Placing Ecuador's indigenous coup of January 2000 in historical perspective. In: Langer, Erick and Munoz, Elena (ed.). Contemporary indigenous movements in Latin America. Wilmington: SR Books, 65–80.
———. 2007. Gender, Indian, nation: The contradictions of making Ecuador, 1830–1925. Tucson: The University of Arizona Press.
O'Neill, Maggie; Giddens, Sara; Breatnach, Patricia; Bagley, Carl; Bourne, Darren; and Judge, Tony. 2002. Renewed methodologies for social research: Ethno-mimesis as performative praxis. In: The Sociological Review 50(1): 69–88.
Oslender, Ulrich. 2007. Revisting the hidden transcript: Oral tradition and black cultural politics in the Colombian Pacific coast region. In: Environment and Planning D: Society and Space 25(6): 1103–1129.
———. 2008. Las políticas de etnicidad en América Latina: Comunidades indígenas y afrodescendientes como nuevos sujetos políticos y el desafío descolonial. In: Cairo Carou, Heriberto and Mignolo, Walter D. (ed.). Las vertientes Americanas del pensamiento y el proyecto des-colonial. El resurgimiento de los pueblos indígenas y afrolatinos como sujetos políticos. Madrid: Trampa, 101–124.
Paasi, Anssi. 2006. Texts and contexts in the globalizing academic marketplace: Comments on the debate on geopolitical remote sensing. In: Eurasian Geography and Economics 47(2): 216–220.
Pacari, Nina. 2005. Case study: Ecuador. Unfinished business. The political participation of indigenous women. In: IDEA (ed.). Women in parliament: Beyond numbers. Stockholm: IDEA, 72–80.
Pain, Rachel. 2009. Globalized fear? Towards an emotional geopolitics. In: Progress in Human Geography 33(4): 466–486.
———. 2010. The new geopolitics of fear. In: Geography Compass 4(3): 226–240.
Pain, Rachel and Francis, Peter. 2003. Reflections on participatory research. In: Area 35(1): 46–54.
Pain, Rachel; Panelli, Ruth; Kindon, Sara; and Little, Jo. 2010. Moments in everyday/distant geopolitics: Young people's fears and hopes. In: Geoforum 41(6): 972–982.
Pain, Rachel and Smith, Susan J. 2008. Fear: Critical geopolitics and everyday life. Aldershot: Ashgate.
Pallares, Amalia. 2002. From peasant struggles to Indian resistance: The Ecuadorian Andes in the late twentieth century. Norman: University of Oklahoma Press.
———. 2007. Contesting membership: Citizenship, pluriculturalism(s), and the contemporary indigenous movement. In: Clark, Kim and Becker, Marc (ed.). Highland Indians and the state in modern Ecuador. Pittsburgh: University of Pittsburgh Press, 139–154.
Palmary, Ingrid. 2011. 'In your experience': Research as gendered cultural translation. In: Gender, Place & Culture: A Journal of Feminist Geography 18(1): 99 – 113.
Parpart, Jane. 1993. Who is the 'Other'? A postmodern feminist critique of women and development theory and practice In: Development and Change 24(3): 439–446.
Pateman, Carole. 1988. The sexual contract. Stanford: Stanford University Press.
Pateman, Carole and Mills, Charles. 2007. Contract and domination. Cambridge: Polity Press.
Paxton, Pamela; Kunovich, Sheri; and Hughes, Melanie. 2007. Gender in politics. In: Annual Review of Sociology 33: 263–84.
Peake, Linda. 1993. 'Race' and sexuality: Challenging the patriarchal structuring of urban social space. In: Environment and Planning D: Society and Space 11(4): 415–432.
Peake, Linda and Trotz, Alissa. 1999. Gender, ethnicity and place: Women and identities in Guayana. London: Routledge.
Peck, Jamie. 2011. Geographies of policy: From transfer-diffusion to mobility-mutation. In: Progress in Human Geography 35(6): 773–797.

Pequeño, Andrea. 2009. Vivir violencia, cruzar los límites. Prácticas y discursos en torno a la violencia contra mujeres en comunidades indígenas de Ecuador. In: Pequeño, Andrea (ed.). Participación y políticas de mujeres indígenas en América Latina. Quito: FLACSO, 147–168.

Peschard, Jacqueline. 2003. The quota system in Latin America: General overview. In: International Institute for Democracy and Electoral Assistance (ed.). The implementation of quotas. Latin American experiences: Quota Workshops Report Series No. 2. Stockholm: IDEA, 20–29.

Pile, Steve. 2010. Emotions and affect in recent human geography. In: Transactions of the Institute of British Geographers 35(1): 5–20.

Pink, Sarah. 2001. Doing visual ethnography: Images, media and representation in research. London: Sage.

———. 2004. Performance, self-representation and narrative: Interviewing with video. In: Pole, Christopher (ed.). Seeing is believing? Approaches to visual research. Oxford: Elsevier, 61–78.

———. 2007. Applied visual anthropology: Social intervention and visual methodologies. In: Pink, Sarah (ed.). Visual interventions. New York: Berghahn Books, 3–29.

———. 2008. Mobilising visual ethnography: Making routes, making place and making images. In: Forum Qualitative Social Research 9(3): Art. 36.

———. 2009. Doing sensory ethnography. London: Sage.

Pitkin, Hanna. 1967. The concept of representation. Berkeley: University of California Press.

Power, Marcus; Mohan, Giles; and Mercer, Claire. 2006. Postcolonial geographies of development: Introduction. In: Singapore Journal of Tropical Geography 27(3): 231–234.

Pratt, Geraldine. 1999. Geographies of identity and difference: Marking boundaries. In: Massey, Doreen; Allen, John; and Sarre, P. (ed.). Human geography today. Cambridge Polity Press, 151–168.

———. 2000a. Research performances. In: Environment and Planning D: Society and Space 18(5): 639-651.

———. 2000b. Performativity. In: Johnston, R. J.; Gregory, Derek; Pratt, Geraldine; and Watt, Michael (ed.). Dictionary of Human Geography. Oxford: Blackwell, 578.

———. 2004. Working feminism. Edinburgh: Edinburg University Press.

———. 2009. Circulating sadness: Witnessing Filipina mothers' stories of family separation. In: Gender, Place & Culture: A Journal of Feminist Geography 16(1): 3–22.

Prieto, Mercedes. 2004. Liberalismo y temor: Imaginando los sujetos indígenas en el Ecuador postcolonial, 1895–1950. Quito: FLACSO.

Prieto, Mercedes; Cuminao, Clorinda; Flores, Alejandra; Maldonado, Gina; and Pequeño, Andrea. 2006. Respcto, discriminación y violencia: Mujeres indígenas en Ecuador, 1990–2004. In: Lebon, Nathalie and Maier, Elizabeth (ed.). De lo privado a lo público: 30 años de lucha ciudadana de las mujeres en América Latina. México: UNIFEM, 158–180.

Prieto, Mercedes and Goetschel, Ana María. 2008. El sufragio femenino en Ecuador, 1884–1940. In: Prieto, Mercedes (ed.). Mujeres y escenarios ciudadanos. Quito: FLACSO, 299–330.

Prieto, Mercedes and Herrera, Gioconda. 2007. Género y nación en América Latina. In: Íconos. Revista de Ciencias Sociales 28: 31–34.

Prieto, Mercedes; Pequeño, Andrea; Cominao, Clorinda; Flores, Alejandra; and Maldonado, Gina. 2010. Respect, discrimination, and violence: Indigenous women in Ecuador 1990–2007. In: Maier, Elizabeth and Lebon, Nathalie (ed.). Women's activism in Latin America and the Caribbean: Engendering social justice, democratizing citizenship. New Brunswick: Rutgers University Press, 203–218.

Pugh, Jonathan. 2005. The disciplinary effects of communicative planning in Soufriere, St Lucia: Governmentality, hegemony and space-time-politics. In: Transactions of the Institute of British Geographers 30(3): 307–321.

———. 2007. Book review: On the political, by Chantal Mouffe. In: Area 39(1): 130–131.

Quezada, Alexandra. 2009. Del derecho al voto a la presencia de las mujeres en la vida política nacional. In: Rodas Morales, Raquel (ed.). Historia del voto feminino en el Ecuador. Quito: CONAMU, 143–207.

Radcliffe, Sarah. 1994. (Representing) post-colonial women: Authority, difference and feminism. In: Area 26(1): 25–32.

———. 1996. Gendered nations: Nostalgia, development and territory in Ecuador. In: Gender, Place & Culture: A Journal of Feminist Geography 3(1): 5–22.

———. 1997. The geographies of indigenous self-representation in Ecuador: Hybridity, gender and resistance. In: European Review of Latin America and Caribbean Studies 63: 9–27.

———. 1999. Reimagining the nation: community, difference, and national identities among indigenous and mestizo provincials in Ecuador. In: Environment and Planning A 31(1): 37-52.

———. 2000. Entangling resistance, ethnicity, gender and nation in Ecuador. In: Sharp, Joanne; Routledge, Paul; Philo, Chris; and Paddison, Ronan (ed.). Entanglements of power: Geographies of domination/resistance. London: Routledge, 164–181.

———. 2002. Indigenous women, rights and the nation state in the Andes. In: Craske, Nikki and Molyneux, Maxine (ed.). Gender and the politics of rights and democracy in Latin America. Basingstoke: Palgrave, 149–172.

———. 2005. Development and geography: Towards a postcolonial development geography? In: Progress in Human Geography 29(3): 291–298.

———. 2006. Development and geography: Gendered subjects in development processes and interventions In: Progress in Human Geography 30(4): 524–532.

———. 2007. Latin American indigenous geographies of fear: Living in the shadow of racism, lack of development, and antiterror measures. In: Annals of the Association of American Geographers 97(2): 385–397.

———. 2008a. Las mujeres indígenas ecuatorianas bajo la gobernabilidad multicultural y de género. In: Wade, Peter; Urrea, Fernando; and Viveros, Mara (ed.). Raza, etnicidad y sexualidades: ciudadanía y multiculturalismo en América Latina. Bogotá: Centro de Estudios Sociales, 105–136.

———. 2008b. Women's movement in twentieth-century Ecuador. In: De la Torre, Carlos and Striffler, Steve (ed.). The Ecuador reader: History, culture, politics. Durham NC: Duke University Press, 284–296.

———. 2010. Epílogo: Historias de vida de mujeres indígenas a través de la educación y el liderazgo. Intersecciones de raza, género y locación. In: Coronel, Valeria and Prieto, Mercedes (ed.). Celebraciones centenarias y negociaciones por la nación ecuatoriana. Quito: FLACSO, 317–349.

———. 2012. Development for a postneoliberal era? Sumak kawsay, living well and the limits to decolonisation in Ecuador. In: Geoforum 43(2): 240-249.

Radcliffe, Sarah and Laurie, Nina. 2006. Indigenous groups, culturally appropriate development, and the socio-spatial fix of Andean development. In: Radcliffe, Sarah (ed.). Culture and development in a globalizing world. London: Routledge, 83–106.

Radcliffe, Sarah; Laurie, Nina; and Andolina, Robert. 2002. Reterritorialised space and ethnic political participation: Indigenous municipalities in Ecuador. In: Space & Polity 6(3): 289–305.

———. 2003. The transnationalization of gender and reimagining Andean indigenous development In: Signs: Journal of Women in Culture and Society 29(2): 387–416.

Radcliffe, Sarah and Pequeño, Andrea. 2010. Ethnicity, development and gender: Tsáchila indigenous women in Ecuador. In: Development and Change 41(6): 983–1016.

Radcliffe, Sarah and Westwood, Sallie. 1996. Remaking the nation. Place, identity and politics in Latin America. London: Routledge.

Raghuram, Parvati and Madge, Clare. 2006. Towards a method for postcolonial development geography? Possibilities and challenges. In: Singapore Journal of Tropical Geography 27(3): 270–288.

Randeria, Shalini. 2002. Entangled histories of uneven modernities: Civil society, caste solidarities and legal pluralism in post-colonial India. In: Elkana, Yehuda (ed.). Unraveling ties. Frankfurt a.M.: Campus, 284–311.

Reason, Peter and Bradbury, Hilary (ed.). 2001. Handbook of action research. Participatory inquiry and practice. London: Sage.

Reel, Monte. 2005. Long fall in Ecuador: Populist to pariah. In Washington Post, p. A13.

República del Ecuador. 2008. Constitution of the Republic of Ecuador. Quito: Official Register 20. October 2008

———. 2009. Normas generales para las elecciones dispuestas en el régimen de transición de la Constitución de la República del Ecuador. Quito: El Consejo Nacional Electoral.

Riaño, Yvonne. 2011. Drawing new boundaries of participation: Experiences and strategies of economic citizenship among skilled migrant women in Switzerland. In: Environment and Planning A 43(7): 1530–1546.

Rice, Roberta. 2011. From the ground up: The challenge of indigenous party consolidation in Latin America. In: Party Politics 17(2): 171–188.

Rice, Roberta and Van Cott, Donna Lee. 2006. The emergence and performance of indigenous peoples' parties in South America: A subnational statistical analysis. In: Comparative Political Studies 39(6): 709–732.

Ríos Tobar, Marcela. 2008. Mujer y política: El impacto de las cuotas de género en América Latina. Santiago de Chile: IDEA International, FLACSO Chile.

Rivera Vélez, Fredy and Ramírez Gallegos, Franklin. 2005. Ecuador: Democracy and economy in crisis. In: Crandall, Russell; Paz, Guadalupe; and Roett, Riordan (ed.). The Andes in focus: Security, democracy and economic reform. London: Lynne Rienner Publishers, 121–149.

Robinson, Jenny. 2000. Feminism and the spaces of transformation. In: Transactions of the Institute of British Geographers 25(3): 285–301.

Roelvink, Gerda. 2010. Collective action and the politics of affect. In: Emotion, Space and Society 3(2): 111–118.

Rose, Gillian. 1993. Feminism and geography. The limits of geographical knowledge. Oxford: Polity Press.

———. 1997a. Engendering the slum: Photography in East London in the 1930s. In: Gender, Place & Culture: A Journal of Feminist Geography 4(3): 277–300.

———. 1997b. Situating knowledges: Positionality, reflexivities and other tactics. In: Progress in Human Geography 21(3): 305–320.

———. 2004. 'Everyone's cuddled up and it just looks really nice': An emotional geography of some mums and their family photos. In: Social & Cultural Geography 5(4): 549–564.

———. 2011. Visual methodologies: An introduction to researching with visual materials London: Sage.

Rose, Mitch. 2002. The seductions of resistance: Power, politics, and a performative style of systems. In: Environment and Planning D: Society & Space 20(4): 383–400.

Rose-Redwood, Reuben. 2008. 'Sixth Avenue is now a memory': Regimes of spatial inscription and the performative limits of the official city-text. In: Political Geography 27(8): 875–894.

Rose-Redwood, Reuben and Alderman, Derek. 2011. Critical interventions in political toponymy. In: ACME 10(1): 1–6.

Ruddick, Susan. 1996. Constructing difference in public spaces: Race, class, and gender as interlocking systems. In: Urban Geography 17(2): 132–151.

Safa, Helen. 2005. Challenging mestizaje: A gender perspective on indigenous and afrodescendant movements in Latin America. In: Critique of Anthropology 25(3): 307–330.

Said, Edward. 1978. Orientalism. New York: Vintage books.

Sanbonmatsu, Kira. 2010. Organizing American politics, organizing gender. In: Leighley, Jan E. (ed.). The Oxford Handbook of American elections and political behavior. Oxford: Oxford University Press, 471–492.

Sánchez-Parga, José. 2007. El movimiento indígena Ecuatoriano. La larga ruta de la comunidad al partido. Quito: CAAP.

Sapiro, Virgina. 1981. When are interests interesting? The problem of political representation of women. In: The American Political Science Review 75(3): 701–716.

Sauer, Birgit. 2008. Formwandel politischer Institutionen im Kontext neoliberaler Globalisierung und die Relevanz der Kategorie Geschlecht. In: Casale, Rita and Rendtorff, Barbara (ed.). Was kommt nach der Genderforschung? Zur Zukunft der feministischen Theoriebildung. Bielefeld: Transcript, 237–254.

Sauer, Carl O. 1918. Geography and the gerrymander. In: American Political Science Review 12(3): 403–426.

Sawyer, Suzana. 1997. The 1992 Indian mobilization in lowland Ecuador. In: Latin American Perspectives 24(3): 65–82.

Schaffner, Brian F. 2006. The political geography of campaign advertising in U.S. House elections. In: Political Geography 25(7): 775–788.

Scharff, Christina. 2011. Towards a pluralist methodological approach: Combining performativity theory, discursive psychology and theories of affect. In: Qualitative Research in Psychology 8(2): 210–221.

Schurr, Carolin. 2009a. Andean rural local governments in-between powerscapes. Eichstätt: Mesa Redonda.

———. 2009b. Feminisierung und Indigenisierung politischer Räume als Prozess kultureller Dekolonialisierung. In: Wastl-Walter, Doris (ed.). Gender Geographien. Geschlecht und Raum als soziale Konstruktionen. Stuttgart: Franz Steiner Verlag 52–54.

———. 2011. Intersektionalität als Konzept der Gender-Geographien. In: Feministisches Geo-RundMail 44.

———. 2012a. Visual ethnography for performative geographies: How women politicians perform identities on Ecuadorian political stages. In: Geographica Helvetica 67(4): 195–202.

———. 2012b. Pensando emoções a partir de uma perspectiva interseccional: As geografias emocionais das campanhas eleitorais equatorianas. In: Revista Latino-Americana de Geografia e Gênero 3(2): 3–15.

———. forthcoming. Emotionen, Affekte und non-repräsentationale Geographien. In: Geographische Zeitschrift.

———. in press. Performativity and antagonism as keystones for a political geography of change. In: Glass, Michael and Rose-Redwood, Reuben (ed.). Performativity, politics and social space. New York: Routledge.

Schurr, Carolin and Fredrich, Bettina. 2011. Feministische Politische Geographie. In: Feministisches Geo-RundMail 48.

Schurr, Carolin and Kaspar, Heidi. 2013. Feminists in the wild - Geschlecht im Feld. In: Feministisches Geo-RundMail 54.

Schurr, Carolin and Segebart, Dörte. 2012. Tackling feminist postcolonial critique through participatory and intersectional approaches. In: Geographica Helvetica 67(3): 147-154.

Schurr, Carolin and Wintzer, Jeannine (ed.). 2012. Geschlecht und Raum feministisch denken. Bern: eFeF.

Scott, James. 1990. Domination and the arts of resistance. Hidden transcript. Yale: Yale University Press.

Seagar, Joni. 1997. The state of women in the world atlas. London: Penguin.

Secor, Anna. 2001. Towards a feminist counter-geopolitics: Gender, space and Islamist politics in Istanbul. In: Space & Polity 5(3): 191–211.

———. 2001. Ideologies in crisis: Political cleavages and electoral politics in Turkey in the 1990s. In: Political Geography 20(5): 539–560.

———. 2003. Belaboring gender: The spatial practice of work and the politics of 'making do' in Istanbul. In: Environment and Planning A 35(12): 2209–2227.

———. 2004. Feminizing electoral geography. In: Staeheli, Lynn; Kofman, Eleonore; and Peake, Linda (ed.). Mapping women, making politics. Feminist perspectives on political geography New York: Routledge, 261–272.

Segebart, Dörte. 2007. Partizipatives Monitoring als Instrument zur Umsetzung von Good Local Governance – Eine Aktionsforschung im östlichen Amazonien/Brasilien. Tübingen: Geographisches Institut der Universität Tübingen.

Segebart, Dörte and Schurr, Carolin. 2010. Was kommt nach Gendermainstreaming? Herausforderungen an die geographische Entwicklungsforschung in der Geschlechterforschung. In: Geographische Rundschau 62(10): 58–63.

Selverston-Scher, Melina. 2001. Ethnopolitics in Ecuador: Indigenous rights and the strengthening of democracy. Miami: University Press of Miami.

Sharp, Joanne. 2003. Deconstructing private/public: Gender in a political and patriarchal world. In: Anderson, Kay; Domosh, Mona; Pile, Steve; and Thrift, Nigel (ed.). The Handbook of Cultural Geography. London: Sage, 473–484.

———. 2004. Doing feminist political geographies. In: Kofman, Elenore; Peake, Linda; and Staeheli, Lynn (ed.). Mapping women, making politics. Feminist perspectives on political geography New York: Routledge, 87–98.

———. 2005. Geography and gender: Feminist methodologies in collaboration and in the field. In: Progress in Human Geography 29(3): 304–309.

———. 2009a. Geographies of postcolonialism. London: Sage.

———. 2009b. Geography and gender: What belongs to feminist geography? Emotion, power and change. In: Progress in Human Geography 33(1): 74–80.

———. 2011. A subaltern critical geopolitics of the war on terror: Postcolonial security in Tanzania. In: Geoforum 42(3): 297–305.

Sharp, Joanne; Routledge, Paul; Philo, Chris; and Paddison, Ronan (ed.). 2000. Entanglements of power: Geographies of domination/resistance. London: Routledge.

Sidaway, James. 2007. Spaces of postdevelopment. In: Progress in Human Geography 31(3): 345–361.

Sidaway, James D. 1992. In other worlds: On the politics of research by 'First world' geographers in the 'Third world'. In: Area 24(4): 403–408.

Siegfried, André. 1913. Tableau politique de la France de l'Ouest sous la Troisième République. Geneva.

Simien, Evelyn. 2007. Doing intersectionality research: From conceptual issues to practical examples. In: Politics & Gender 3(02): 264–271.

Simonsen, Kirsten. 2007. Practice, spatiality and embodied emotions: An outline of a geography of practice. In: Human Affairs 17(2): 168–181.

———. 2013. In quest of a new humanism: Embodiment, experience and phenomenology as critical geography. In: Progress in Human Geography 37(1): 10–26.

Simpson, Paul. 2011. 'So, as you can see...': Some reflections on the utility of video methodologies in the study of embodied practices. In: Area 43(3): 343–352.

Slater, David. 2002. Other domains of democratic theory: Space, power, and the politics of democratization. In: Environment and Planning D: Society and Space 20(3): 255–276.

———. 2008. Power and social movements in the other Occident: Latin America in an international context. In: Stahler-Sholk, Richard; Vanden, Harry; and Kuecker, Glen David (ed.). Latin American social movements in the twenty-first century: Resistance, power, and democracy. Lanham, Maryland: Rowman & Littlefield Publishers, 17–38.

Slocum, Rachel. 2008. Thinking race through corporeal feminist theory: Divisions and intimacies at the Minneapolis farmers' market. In: Social & Cultural Geography 9(8): 849–869.

Sloterdijk, Peter. 2008. Zorn und Zeit. Frankfurt a.M.: Suhrkamp.

Smith, Mick; Davidson, Joyce; Cameron, Deborah; and Bondi, Liz (ed.). 2009a. Emotion, place and culture. Farnham: Ashgate.

———. 2009b. Introduction: Geography and emotion – emerging constellations. In: Smith, Mick; Davidson, Joyce; Cameron, Deborah; and Bondi, Liz (ed.). Emotion, place and culture. Farnham: Ashgate, 1–20.

Smith, Sara. 2011. 'She says herself, "I have no future"': Love, fate and territory in Leh District, India. In: Gender, Place & Culture: A Journal of Feminist Geography 18(4): 455–476.

———. 2012. Intimate geopolitics: Religion, marriage, and reproductive bodies in Leh, Ladakh. In: Annals of the Association of American Geographers.

Smooth, Wendy. 2006. Intersectionality in electoral politics: A mess worth making. In: Politics & Gender 2(3): 400–414.

Sosa-Buchholz. 2010. Changing images of male and female in Ecuador: José María Velasco Ibarra and Abdalá Bucaram. In: Kampwirth, Karen (ed.). Gender and populism in Latin America: Passionate politics. Pennsylvania: The Pennsylvania State University Press, 47–66.

Sparke, Matthew. 2007. Geopolitical fears, geoeconomic hopes, and the responsibilities of geography. In: Annals of the Association of American Geographers 97(2): 338–349.

Spencer, Jonathan. 2007. Anthropology, politics, and the state: Democracy and violence in South Asia. Cambridge: Cambridge University Press.

Spencer, Stephen. 2011. Visual research methods in the social sciences. London: Routledge.

Spivak, Gayatri Chakravorty. 1988. Can the subaltern speak? In: Nelson, Cary and Grossberg, Lawrence (ed.). Marxism and the interpretation of culture. Urbana, IL: University of Illinois Press, 271–313.

———. 1990. The postcolonial critic: Interviews, strategies, dialogues. New York: Routledge.

Staeheli, Lynn. 1996. Publicity, privacy, and women's political action. In: Environment and Planning D: Society and Space 14(5): 601–619.

———. 2004. Mobilizing women, mobilizing gender: Is it mobilizing difference? In: Gender, Place & Culture: A Journal of Feminist Geography 11(3): 347–372.

———. 2008. Political geography: Difference, recognition, and the contested terrains of political claims-making. In: Progress in Human Geography 32(4): 561–570.

———. 2011. Political geography: Where's citizenship? In: Progress in Human Geography 35(3): 393–400.

Staeheli, Lynn and Kofman, Eleonore. 2004a. Mapping gender, making politics: Toward feminist political geographies. In: Staeheli, Lynn; Kofman, Eleonore; and Peake, Linda (ed.). Mapping women, making politics. Feminist perspectives on political geography New York: Routledge, 1–13

Staeheli, Lynn; Kofman, Eleonore; and Peake, Linda (ed.). 2004b. Mapping women, making politics. Feminist perspective on political geography. New York: Routledge.

Staeheli, Lynn and Lawson, Victoria. 1994. A discussion of 'women in the field': The politics of feminist fieldwork. In: The Professional Geographer 46(1): 96–102.

Staeheli, Lynn; Ledwith, Valerie; Ormond, Meghann; Reed, Katie; Sumpter, Amy; and Trudeau, Daniel. 2002. Immigration, the internet, and spaces of politics. In: Political Geography 21(8): 989–1012.

Staeheli, Lynn and Mitchell, Don. 2004. Spaces of public and private: Locating politics. In: Barnett, Clive and Low, Murray (ed.). Spaces of democracy: Geographical perspectives on citizenship, participation and representation. London: Sage, 147–160.

Stevens, Evelyn. 1973. Marianismo: The other face of Machismo in Latin America. In: Pescatelo, Ann (ed.). Female and male in Latin America. Pittsburgh: University of Pittsburgh Press, 90–101.

Stokes-Brown, Atiya Kai and Neal, Melissa Olivia. 2008. Give 'em something to talk about: The influence of female candidates' campaign issues on political proselytizing. In: Politics & Policy 36(1): 32–59.

Stolcke, Verena. 1993. Is sex to gender as race is to ethnicity? In: del Valle, T. (ed.). Gendered anthropology. London: Routledge, 17–38.

Strüver, Anke. 2005a. Macht Körper Wissen Raum? Ansätze für eine Geographie der Differenzen. Wien: Universität Wien.

———. 2005b. Stories of the 'boring border': The Dutch-German boderscape in people's minds. Münster: Lit Verlag.

Strüver, Anke and Wucherpfennig, Claudia. 2009. Performativität. In: Glasze, Georg and Mattissek, Annika (ed.). Handbuch Diskurs und Raum. Theorien und Methoden für die Humangeographie sowie die sozial- und kulturwissenschaftliche Raumforschung. Bielefeld: Transcript, 107–127.

Sturken, Marita and Cartwright, Lisa. 2009. Practices of looking: An introduction to visual culture. Oxford: Oxford University Press.

Stutzman, Robert. 1981. El mestizaje: An all-inclusive ideology of exclusion. In: Whitten, Norman (ed.). Cultural transformations and ethnicity in modern Ecuador. Chicago: University of Illinois Press, 45–95.

Sundberg, Juanita. 2005. Looking for the critical geographer, or why bodies and geographies matter to the emergence of critical geographies of Latin America. In: Geoforum 36(1): 17–28.

Swyngedouw, Erik 2008. Where is the political? London: Antipode Lecture, IBG/RGS annual conference 2007.

Taylor, Diana. 2003. Bush's happy performative. In: The Drama Review 47(3): 5–8.

Taylor, Peter. 1978. Progress report: Political geography. In: Progress in Human Geography 2(1): 153–162.

Taylor, Peter and Flint, Colin. 2000. Political geography. World-economy, nation-state and locality. Harlow: Pearson Education.

Taylor, Peter and Johnston, Ron. 1979. Geography of elections. London: Penguin.

Taylor, Verta and Rupp, Leila. 2002. Loving internationalism: The emotion culture of transnational women's organizations, 1888–1945. In: Mobilization 7(2): 141–158.

Thien, Deborah. 2005. After or beyond feeling? A consideration of affect and emotion in geography. In: Area 37(4): 450–454.

———. 2007. Disenchanting democracy. In: Area 39(1): 134–135.

———. 2011. Emotional life. In: Del Casino Jr., Vincent; Thomas, Mary E.; Cloke, Paul; and Panelli, Ruth (ed.). A Companion to Social Geography. Oxford: Blackwell, 309–325.

Thomas, Mary E. 2008. Resisting mothers, making gender: Teenage girls in the United States and the articulation of femininity. In: Gender, Place & Culture: A Journal of Feminist Geography 15(1): 61–74.

Thomassen, Lasse. 2005. Reading radical democracy: A commentary on Clive Barnett. In: Political Geography 24(5): 631–639.

Thrift, Nigel. 1997. The still point: Resistance, expressive embodiment and dance. In: Pile, Steve and Keith, Michael (ed.). Geographies of resistance. London: Routledge, 124–151.

———. 2000. It's the little things. In: Dodds, Klaus and Atkinson, D. (ed.). Geopolitical traditions: A century of geopolitical thought. London: Routledge, 380–387.

———. 2003. Performance and…. In: Environment and Planning A 35(11): 2019–2024.

———. 2004. Intensities of feeling: Towards a spatial politics of affect. In: Geografiska Annaler 86(1): 57–78.

———. 2006. Space, place and time. In: Goodin, Robert E. and Tilly, Charles (ed.). The Oxford handbook of contexual political analysis. Oxford: Oxford University Press, 547–563.

———. 2008. Non-representational theory: Space, politics, affect. London: Routledge.

———. 2009. Understanding the affective spaces of political performance. In: Smith, Mick; Davidson, Joyce; Cameron, Deborah; and Bondi, Liz (ed.). Emotion, place and culture. Farnham: Ashgate, 79–96.

Thrift, Nigel and Dewsbury, John-David. 2000. Dead geographies – and how to make them live. In: Environment and Planning D: Society & Space 18(4): 411–432.

Tibán, Lourdes. 2005. Las mujeres y la participación en la equidad en las organizaciones indígenas de Ecuador. In: Sánchez Nestór, Martha (ed.). La doble mirada. Voces e historias de mujeres indígenas latinoamericanas. Mexico City: Instituto de Liderazgo Simone de Beauvoir & Unifem, 51–61.

Tolia-Kelly, Divya. 2006. Affect – an ethnocentric encounter? Exploring the 'universalist' imperative of emotional/affectual geographies. In: Area 38(2): 213–217.

Townsend-Bell, Erica. 2011. Intersectional advances? Inclusionary and intersectional state action in Uruguay. In: APSA 2011 Annual Meeting Paper.

Tuaza, Luis Alberto. 2009. Cansancio organizativo. In: Martínez, Carmen (ed.). Repensando los movimientos indígenas. Quito: FLACSO, 123–146.

Tuhiwai Smith, Linda. 1999. Decolonizing methodologies: Research and indigenous peoples. London: Zed Books.

Tully, James. 2000. The struggles of indigenous peoples for and of freedom. In: Ivison, Duncan; Patton, Paul; and Sanders, Will (ed.). Political theory and the rights of indigenous peoples. New York: Cambridge University Press, 36–59.

UNDP. 2002. Common country assessment: Ecuador. Visión del sistema de Naciones Unidas sobre la situación del Ecuador. New York: UNDP.

United Nations. 2001. Background briefing on intersectionality, Working Group on Women and Human rights, 45th session of the UN CSW.

Valdivia, Gabriela. 2005. On indigeneity, change, and representation in the northeastern Ecuadorian Amazon. In: Environment and Planning A 37(2): 285–303.

———. 2009. Indigenous bodies, indigenous minds? Towards an understanding of indigeneity in the Ecuadorian Amazon. In: Gender, Place & Culture: A Journal of Feminist Geography 16(5): 535–551.

Valentine, Gill. 1989. The geography of women's fear. In: Area 21(4): 385–90.

———. 2002. People like us: Negotiating sameness and difference in the research process. In: Moss, Pamela (ed.). Feminist geography in practice: Research and methods. Oxford: Blackwell, 116–126.

———. 2007. Theorizing and researching intersectionality: A challenge for feminist geography. In: The Professional Geographer 59(1): 10–21.

———. 2008. Living with difference: Reflections on geographies of encounter. In: Progress in Human Geography 32(3): 323–337.

Van Cott, Donna Lee. 2000. Party system development and indigenous populations in Latin America: The Bolivian case. In: Party Politics 6(2): 155–174.

———. 2006. Turning crisis into opportunity: Achievements of excluded groups in the Andes. In: Drake, Paul W. and Hershberg, Eric (ed.). State and society in conflict: Comparative perspectives on Andean crisis. Pittsburgh: University of Pittsburgh Press, 157–188.

———. 2008. Radical democracy in the Andes. Cambridge: Cambridge University Press.

van der Hoogte, Liesbeth and Kingma, Koos. 2004. Promoting cultural diversity and the rights of women: The dilemmas of 'intersectionality' for development organisations. In: Gender & Development 12(1): 47–55.

Van Zoonen, Liesbet. 2005. Entertaining the citizen: When politics and popular culture converge. Oxford: Rowman & Littlefield Publishers.

Vega Ugalde, Silvia. 2004. La cuota electoral de las mujeres: Elementos para un balance. In: Cañete, María Fernanda (ed.). Reflexiones sobre mujer y política. Memoria del seminario nacional 'Los cambios políticos en el Ecuador: Perspectivas y retos para las mujeres'. Quito: Abya-Yala, 43–58.

———. 2005. La cuota electoral en Ecuador: Nadando a contracorriente en un horizonte esperanzador. In: León, Magdalena (ed.). Nadando contra la corriente. Mujeres y cuotas políticas en los páises Andinos. Quito: UNIFEM, UNFPA, FLACSO 169–206.

Verne, Julia. 2012. Ethnographie und ihre Folgen für die Kulturgeographie: Eine Kritik des Netzwerkkonzepts in Studien zu translokaler Mobilität. In: Geographica Helvetica 67(4): 185–194.

Vilas, Carlos. 1998. Buscando al leviatán: Hipótesis sobre ciudadanía, desigualdad, y democracia. In: Sader, Emir (ed.). Democracia sin exclusiones ni excluidos. Caracas: Nueva Sociedad, 115–135.
Walgenbach, Katharina; Dietze, Gabriele; Hornscheidt, Antje; and Palm, Kerstin. 2007. Gender als interdependente Kategorie. Neue Perspektive auf Intersektionalität, Diversität und Heterogenität. Opladen: Verlag Barbara Budrich.
Walsh, Catherine. 2003. Políticas (inter)culturales y gobiernos locales: Experiencias ecuatorianas. In: Mayor, IDCT/Alcaldía (ed.). Políticas culturales urbanas: Experiencias Europeas y Americanas. Bogotá: IDCT/Alcaldía, 110–119.
———. 2008a. Interculturalidad, plurinacionalidad y decolonialidad: Las insurgencias político-epistémicas de refundar el Estado. In: Tabula Rasa 9(2): 131–152.
———. 2008b. (Post)coloniality in Ecuador: The indigenous movement's practices and politics of (re)signification and decolonization. In: Morana, Mabel; Dussel, Enrique; and Jáuregui, Carlos (ed.). Coloniality at large: Latin America and the postcolonial debate. Durham: Duke University Press, 506–518.
———. 2009. The plurinational and intercultural state: Decolonization and State re-founding in Ecuador. In: Kult 6 – Special issue: Epistemologies of transformation: The Latin American decolonial option and its ramifications Fall 2009: 65–84.
Warf, Barney. 2011. Class, ethnicity, religion and place in the 2008 US presidential election. In: Warf, Barney and Leib, Jonathan (ed.). Revitalizing electoral geography. Farnham: Ashgate, 133–155.
Warf, Barney and Leib, Jonathan (ed.). 2011. Revitalizing electoral geography. Farnham: Ashgate.
West, Candace and Zimmerman, Don. 1987. Doing gender. In: Gender & Society 1(2): 125–151.
West, Traci. 1999. Wounds of the spirit: Black women, violence, and resistance ethics. New York: New York University Press.
West-Newman, Lane. 2004. Anger, ethnicity, and claiming rights. In: ethnicities 4(1): 27–52.
Whitten, Norman E.; Scott Whitten, Dorothea; and Chango, Alfonso. 1997. Return of the Yumbo: the indigenous Caminata from Amazonia to Andean Quito. In: American Ethnologist 24(2): 355–391.
Wiles, Janine L.; Rosenberg, Mark W.; and Kearns, Robin A. 2005. Narrative analysis as a strategy for understanding interview talk in geographic research. In: Area 37(1): 89–99.
Wilson, Chris. 1997. The myth of Santa Fe: Creating a modern regional tradition. Albuquerque, NM: University of New Mexico Press.
Winker, Gabriele and Degele, Nina. 2009. Intersektionalität. Zur Analyse sozialer Ungleichheit. Bielefeld: Transcript.
———. 2011. Intersectionality as multi-level analysis: Dealing with social inequality. In: European Journal of Women's Studies 18(1): 51–66.
Wood, Nichola and Smith, Susan J. 2004. Instrumental routes to emotional geographies. In: Social & Cultural Geography 5(4): 533–548.
Woodward, Keith. 2011. Affective life. In: Del Casino Jr., Vincent; Thomas, Mary E.; Cloke, Paul; and Panelli, Ruth (ed.). A Companion to Social Geography. Oxford: Blackwell, 326–345.
Woodward, Keith and Lea, Jennifer. 2010. Geographies of affect. In: Smith, Susan; Pain, Rachel; Marston, Sallie; and Jones III, John Paul (ed.). The Sage Handbook of Social Geographies. London: Sage, 154–175.
Woon, Chih Yuan. 2013. For 'emotional fieldwork' in critical geopolitical research on violence and terrorism. In: Political Geography 33(1): 31–41.
Wright, Sarah. 2008. Practising hope: Learning from social movement strategies in the Philippines. In: Pain, Rachel and Smith, Susan (ed.). Fear: Critical geopolitics and everyday life. Aldershot: Ashgate, 223–234.

Yang, Guobin. 2000. Achieving emotions in collective action: Emotional processes and movement mobilization in the 1989 Chinese student movement. In: The Sociological Quarterly 41(4): 593–614.

Yashar, Deborah. 1999. Democracy, indigenous movements and the postliberal challenge in Latin America. In: World Politics 52(1): 76–104.

———. 2005. Contesting citizenship in Latin America. The rise of indigenous movements and the postliberal challenge. Cambridge: Cambridge University Press.

———. 2006a. Indigenous politics in the Andes: Changing patterns of recognition, reform, and representation. In: Mainwaring, Scott; Bejarano, Ana María; and Pizarro Leóngomez, Eduardo (ed.). The crisis of democratic representation in the Andes. Stanford: Stanford University Press, 257–294.

———. 2006b. Ethnic politics and political instability in the Andes. In: Drake, Paul W. and Hershberg, Eric (ed.). State and society in conflict: Comparative perspectives on Andean crisis. Pittsburgh: University of Pittsburgh Press, 189–219.

Young, Iris Marion. 1989. Polity and group difference: A critique of the ideal of universal citizenship. In: Ethics 99(2): 250–274.

Young, Lorraine and Barrett, Hazel. 2001. Adapting visual methods: Action research with Kampala street children. In: Area 33(2): 141–152.

Young, Robert. 1995. Colonial desire. Hybridity in theory, culture and race. London: Routledge.

Yuval-Davis, Nira. 2006. Intersectionality and feminist politics. In: European Journal of Women's Studies 13(3): 193–209.

Zamosc, Leon. 2004. The Indian movement in Ecuador: From politics of influence to politics of power. In: Grey Postero, Nancy and Zamosc, Leon (ed.). The struggle for indigenous rights in Latin America. Brighton: Sussex Academic Press, 131–157.

Zapata Galindo, Martha. 2013. Intersektionalität und Gender Studies in Lateinamerika. In: Querelles. Jahrbuch für Frauen- und Geschlechterforschung 16.

Zepada, Beatriz. 2010. Construyendo la nación en el siglo XXI: La 'Patria' en el discurso del presidente Correa. In: Burbano de Lara, Felipe (ed.). Transiciones y rupturas. El Ecuador en la segunda mitad del siglo XX. Quito: FLACSO, 159–193.

ERDKUNDLICHES WISSEN
Schriftenreihe für Forschung und Praxis

Begründet von Emil Meynen.
Herausgegeben von Martin Coy, Anton Escher und Thomas Krings.

Franz Steiner Verlag ISSN 0425–1741

108. Hans-Georg Möller
Tourismus und Regionalentwicklung im mediterranen Südfrankreich
Sektorale und regionale Entwicklungseffekte des Tourismus – ihre Möglichkeiten und Grenzen am Beispiel von Côte d'Azur, Provence und Languedoc-Roussillon
1992. XIV, 413 S. mit 60 Abb., kt.
ISBN 978-3-515-05632-8

109. Klaus Frantz
Die Indianerreservationen in den USA
Aspekte der territorialen Entwicklung und des sozio-ökonomischen Wandels
1993. 298 S. und 20 Taf., kt.
ISBN 978-3-515-06217-6

110. Hans-Jürgen Nitz (Hg.)
The Early Modern World-System in Geographical Perspective
1993. XII, 403 S. mit 67 Abb., kt.
ISBN 978-3-515-06094-3

111. Eckart Ehlers / Thomas Krafft (Hg.)
Shâhjahânâbâd / Old Delhi
Islamic Tradition and Colonial Change
1993. 106 S. mit 14 Abb., 1 fbg. Frontispiz und 1 mehrfbg. Faltkt., kt.
ISBN 978-3-515-06218-3

112. Ulrich Schweinfurth (Hg.)
Neue Forschungen im Himalaya
1993. 293 S. mit 50 Abb., 1 Diagr., 6 Ktn., 35 Fotos, kt.
ISBN 978-3-515-06263-3

113. Rüdiger Mäckel / Dierk Walther
Naturpotential und Landdegradierung in den Trockengebieten Kenias
1993. 309 S. mit 66 Abb., 49 Tab., 36 Fotos (davon 4 fbg.), kt.
ISBN 978-3-515-06197-1

114. Jürgen Schmude
Geförderte Unternehmensgründungen in Baden-Württemberg
Eine Analyse der regionalen Unterschiede des Existenzgründungsgeschehens am Beispiel des Eigenkapitalhilfe-Programms (1979 bis 1989)
1994. XVII, 246 S. mit 13 Abb., 38 Tab., 21 Ktn., kt.
ISBN 978-3-515-06448-4

115. Werner Fricke / Jürgen Schweikart (Hg.)
Krankheit und Raum
Dem Pionier der Geomedizin Helmut Jusatz zum Gedenken
1995. VIII, 254 S. mit 46 Abb. und 1 Taf., kt.
ISBN 978-3-515-06648-8

116. Benno Werlen
Sozialgeographie alltäglicher Regionalisierungen. Bd. 1
Zur Ontologie von Gesellschaft und Raum
1995. X, 262 S., kt.
ISBN 978-3-515-06606-8

117. Winfried Schenk
Waldnutzung, Waldzustand und regionale Entwicklung in vorindustrieller Zeit im mittleren Deutschland
1995. 326 S. mit 48 Tab., 65 Fig., kt.
ISBN 978-3-515-06489-7

118. Fred Scholz
Nomadismus
Theorie und Wandel einer sozio-ökologischen Kulturweise
1995. 300 S. mit 30 Abb., 41 Fotos und 3 fbg. Beilagen, kt.
ISBN 978-3-515-06733-1

119. Benno Werlen
Sozialgeographie alltäglicher Regionalisierungen. Bd. 2
Globalisierung, Region und Regionalisierung
1997. XI, 464 S., kt.
ISBN 978-3-515-06607-5

120. Peter Jüngst
Psychodynamik und Stadtgestaltung
Zum Wandel präsentativer Symbolik und Territorialität von der Moderne zur Postmoderne
1995. 175 S. mit 12 Abb., kt.
ISBN 978-3-515-06534-4

121. Benno Werlen (Hg.)
Sozialgeographie alltäglicher Regionalisierungen. Bd. 3

Ausgangspunkte und Befunde empirischer Forschung
2007. 336 S. mit 28 Abb., 5 Tab., kt.
ISBN 978-3-515-07175-8

122. Zoltán Cséfalvay
Aufholen durch regionale Differenzierung?
Von der Plan- zur Marktwirtschaft – Ostdeutschland und Ungarn im Vergleich
1997. XIII, 235 S., kt.
ISBN 978-3-515-07125-3

123. Hiltrud Herbers
Arbeit und Ernährung in Yasin
Aspekte des Produktions-Reproduktions-Zusammenhangs in einem Hochgebirgstal Nordpakistans
1998. 295 S. mit 40 Abb., 45 Tab. und 8 Taf., kt.
ISBN 978-3-515-07111-6

124. Manfred Nutz
Stadtentwicklung in Umbruchsituationen
Wiederaufbau und Wiedervereinigung als Streßfaktoren der Entwicklung ostdeutscher Mittelstädte, ein Raum-Zeit-Vergleich mit Westdeutschland
1998. 242 S. mit 37 Abb., 7 Tab., kt.
ISBN 978-3-515-07202-1

125. Ernst Giese / Gundula Bahro / Dirk Betke
Umweltzerstörungen in Trockengebieten Zentralasiens (West- und Ost-Turkestan)
Ursachen, Auswirkungen, Maßnahmen
1998. 189 S. mit 39 Abb. und 4 fbg. Kartenbeil., kt.
ISBN 978-3-515-07374-5

126. Rainer Vollmar
Anaheim – Utopia Americana
Vom Weinland zum Walt Disneyland. Eine Stadtbiographie
1998. 289 S. mit 164 Abb., kt.
ISBN 978-3-515-07308-0

127. Detlef Müller-Mahn
Fellachendörfer
Sozialgeographischer Wandel im ländlichen Ägypten
1999. XVIII, 302 S. mit 59 Abb., 31 Fotos und 6 Farbktn., kt.
ISBN 978-3-515-07412-4

128. Klaus Zehner
„Enterprise Zones" in Großbritannien
Eine geographische Untersuchung zu Raumstruktur und Raumwirksamkeit eines innovativen Instruments der Wirtschaftsförderungs- und Stadtentwicklungspolitik in der Thatcher-Ära
1999. 256 S. mit 14 Abb., 31 Tab., 14 Ktn., kt.
ISBN 978-3-515-07555-8

129. Peter Lindner
Räume und Regeln unternehmerischen Handelns
Industrieentwicklung in Palästina aus institutionenorientierter Perspektive
1999. XV, 280 S. mit 33 Abb., 11 Tab. und 1 Kartenbeilage, kt.
ISBN 978-3-515-07518-3

130. Peter Meusburger (Hg.)
Handlungszentrierte Sozialgeographie
Benno Werlens Entwurf in kritischer Diskussion
1999. 269 S., kt.
ISBN 978-3-515-07613-5

131. Paul Reuber
Raumbezogene Politische Konflikte
Geographische Konfliktforschung am Beispiel von Gemeindegebietsreformen
1999. 370 S. mit 54 Abb., kt.
ISBN 978-3-515-07605-0

132. Eckart Ehlers / Hermann Kreutzmann (Hg.)
High Mountain Pastoralism in Northern Pakistan
2000. 211 S. mit 36 Abb., 20 Fotos, kt.
ISBN 978-3-515-07662-3

133. Josef Birkenhauer
Traditionslinien und Denkfiguren
Zur Ideengeschichte der sogenannten Klassischen Geographie in Deutschland
2001. 118 S., kt.
ISBN 978-3-515-07919-8

134. Carmella Pfaffenbach
Die Transformation des Handelns
Erwerbsbiographien in Westpendlergemeinden Südthüringens
2002. XII, 240 S. mit 28 Abb., 12 Fotos, kt.
ISBN 978-3-515-08222-8

135. Peter Meusburger / Thomas Schwan (Hg.)
Humanökologie
Ansätze zur Überwindung der Natur-Kultur-Dichotomie
2003. IV, 342 S. mit 30 Abb., kt.
ISBN 978-3-515-08377-5

136. Alexandra Budke / Detlef Kanwischer / Andreas Pott (Hg.)
Internetgeographien
Beobachtungen zum Verhältnis von Internet, Raum und Gesellschaft
2004. 200 S. mit 28 Abb., kt.

137. Britta Klagge
Armut in westdeutschen Städten
Strukturen und Trends aus stadtteilorientierter Perspektive – eine vergleichende Langzeitstudie der Städte Düsseldorf, Essen, Frankfurt, Hannover und Stuttgart
2005. 310 S. mit 32 s/w- und 16 fbg. Abb., 53 Tab., kt.
ISBN 978-3-515-08556-4

138. Caroline Kramer
Zeit für Mobilität
Räumliche Disparitäten der individuellen Zeitverwendung für Mobilität in Deutschland
2005. XVII, 445 S. mit 120 Abb., kt.
ISBN 978-3-515-08630-1

139. Frank Meyer
Die Städte der vier Kulturen
Eine Geographie der Zugehörigkeit und Ausgrenzung am Beispiel von Ceuta und Melilla (Spanien / Nordafrika)
2005. XII, 318 S. mit 6 Abb., 12 Tab., 3 Farbktn., kt.
ISBN 978-3-515-08602-8

140. Michael Flitner
Lärm an der Grenze
Fluglärm und Umweltgerechtigkeit am Beispiel des binationalen Flughafens Basel-Mulhouse
2007. 238 S. mit 8 s/w-Abb. und 4 Farbtaf., kt.
ISBN 978-3-515-08485-7

141. Felicitas Hillmann
Migration als räumliche Definitionsmacht
2007. 321 S. mit 12 Abb., 18 Tab., 3 s/w- und 5 Farbktn., kt.
ISBN 978-3-515-08931-9

142. Hellmut Fröhlich
Das neue Bild der Stadt
Filmische Stadtbilder und alltägliche Raumvorstellungen im Dialog
2007. 389 S. mit 85 Abb., kt.
ISBN 978-3-515-09036-0

143. Jürgen Hartwig
Die Vermarktung der Taiga
Die Politische Ökologie der Nutzung von Nicht-Holz-Waldprodukten und Bodenschätzen in der Mongolei
2007. XII, 435 S. mit 54 Abb., 31 Tab., 22 Ktn., 92 z.T. fbg. Fotos, geb.
ISBN 978-3-515-09037-7

144. Karl Martin Born
Die Dynamik der Eigentumsverhältnisse in Ostdeutschland seit 1945
Ein Beitrag zum rechtsgeographischen Ansatz
2007. XI, 369 S. mit 78 Abb., 39 Tab., kt.
ISBN 978-3-515-09087-2

145. Heike Egner
Gesellschaft, Mensch, Umwelt – beobachtet
Ein Beitrag zur Theorie der Geographie
2008. 208 S. mit 8 Abb., 1 Tab., kt.
ISBN 978-3-515-09275-3

146. *in Vorbereitung*

147. Heike Egner, Andreas Pott
Geographische Risikoforschung
Zur Konstruktion verräumlichter Risiken und Sicherheiten
2010. XI, 242 S. mit 16 Abb., 3 Tab., kt.
ISBN 978-3-515-09427-6

148. Torsten Wißmann
Raum zur Identitätskonstruktion des Eigenen
2011. 204 S., kt.
ISBN 978-3-515-09789-5

149. Thomas M. Schmitt
Cultural Governance
Zur Kulturgeographie des UNESCO-Welterberegimes
2011. 452 S. mit 60 z.T. farb. Abb., 17 Tab., kt.
ISBN 978-3-515-09861-8

150. Julia Verne
Living Translocality
Space, Culture and Economy in Contemporary Swahili Trade
2012. XII, 262 S. mit 45 Abb., kt.
ISBN 978-3-515-10094-6

151. Kirsten von Elverfeldt
Systemtheorie in der Geomorphologie
Problemfelder, erkenntnistheoretische Konsequenzen und praktische Implikationen
2012. 168 S. mit 13 Abb., kt.
ISBN 978-3-515-10131-8

152. Carolin Schurr
Performing Politics, Making Space
A Visual Ethnography of Political Change in Ecuador
2013. 215 S. mit 36 Abb., 2 Ktn. und 10 Tab., kt.
ISBN 978-3-515-10466-1